The fossils of the Hunsrück Slate
Marine life in the Devonian

This beautifully illustrated book describes one of the most famous fossil deposits known: the Hunsrück Slate of Germany. These spectacular fossils, in which not just the hard parts but also the soft tissues of the animals are preserved in pyrite in many cases, provide the most complete record available of life in the Devonian seas. Among the highlights are trilobites, crinoids and starfish, but rarer forms such as worms and fish are also present.

This book provides the first comprehensive account of these remarkable fossils to be published in English. It is written in an accessible style, and extensively illustrated with photographs and x-radiographs of many of the finest specimens. An up-to-date review of the different plant and animal groups is complemented by accounts of the geological setting, the slate-mining industry, and the preservation and paleoecology. A complete taxonomic list and comprehensive bibliography are included. The book will be of most value to researchers and graduate students in paleontology, geology and evolutionary biology, but it will also be of interest to amateur collectors and natural historians.

T0213068

Cambridge Paleobiology Series

Series editors
D.E.G. Briggs *University of Bristol*
P. Dodson *University of Pennsylvania*
B. J. Macfadden *University of Florida*
J. J. Sepkoski *University of Chicago*
R. A. Spicer *Open University*

Cambridge Paleobiology Series is a new collection of books in the multidisciplinary area of modern paleobiology. The series will provide accessible and readable reviews of the exciting and topical aspects of paleobiology. The books will be written to appeal to advanced students and to professional earth scientists, paleontologists and biologists who wish to learn more about the developments in the subject.

Books in the Series:
1. *The Enigma of Angiosperm Origins* Norman F. Hughes
2. *Patterns and Processes of Vertebrate Evolution* Robert L. Carroll
3. *The Fossils of the Hunsrück Slate: Marine Life in the Devonian* Christoph Bartels, Derek E. G. Briggs and Günther Brassel

The fossils of the Hunsrück Slate

Marine life in the Devonian

CHRISTOPH BARTELS
German Mining Museum, Bochum

DEREK E. G. BRIGGS
Department of Geology, University of Bristol

GÜNTHER BRASSEL
retired

CAMBRIDGE
UNIVERSITY PRESS

CAMBRIDGE UNIVERSITY PRESS
Cambridge, New York, Melbourne, Madrid, Cape Town, Singapore, São Paulo, Delhi

Cambridge University Press
The Edinburgh Building, Cambridge CB2 8RU, UK

Published in the United States of America by Cambridge University Press, New York

www.cambridge.org
Information on this title: www.cambridge.org/9780521117074

First published 1998
This digitally printed version 2009

A catalogue record for this publication is available from the British Library

Library of Congress Cataloguing in Publication data

Bartels, Christoph, 1949–
    The fossils of the Hunsrück Slate: marine life in the Devonian /
Christoph Bartels, Derek E. G. Briggs, Günther Brassel.
        p.   cm. – (Cambridge paleobiology series ; 3)
     ISBN 0 521 44190 0
     1. Paleontology–Devonian.   2. Marine animals, Fossil–Germany–
Hunsrück.   3. Plants, Fossil–Germany–Hunsrück.   4. Hunsrück Slate
(Germany)   I. Briggs, Derek, E. G.   II. Brassel, Günther.
III. Title.   IV. Series.
QE728.B37 1998
560'.174'0943–dc21   97–11979 CIP

ISBN 978-0-521-44190-2 hardback
ISBN 978-0-521-11707-4 paperback

Contents

Introduction

The roof slates of the Hunsrück, Taunus and Eifel regions of western Germany, which form part of the Lower Devonian Hunsrück Slate of the Rhenish Massif, include some of the most important fossil deposits in Europe. They rank alongside other famous German conservation deposits (Konservat-Lagerstätten *sensu* Seilacher 1970) including Odernheim in the Lower Permian of the Saar–Nahe Basin, Holzmaden in the Lower Jurassic of Württemburg, Solnhofen in the Upper Jurassic of Bavaria, and the Eocene occurrences at Messel in Hassia and Geiseltal in Saxony near Halle. However, the preservation of the Hunsrück Slate fossils, which show pyritized soft tissues, differs fundamentally from that in these other deposits.

The Hunsrück Slate fossils are not particularly remarkable in size or spectacular in appearance, and their appeal to the non-specialist is limited. They require skilled and time-consuming preparation. The quality of the preservation can only be appreciated when details are revealed with the aid of a lens or binocular microscope. Appropriate illumination in a particular direction is required to show the fossils to best effect. The results are evident in the photographs in this book, particularly the new ones that are published here for the first time.

The fossils of the Hunsrück Slate did not attract scientific interest before the middle of the nineteenth century, when Ferdinand Roemer (1862) published the first paper on this subject (Fig. 1). Prior to the publication in 1990 of the German book upon which this one is based, only Opitz (1932) had summarized research on the Hunsrück Slate. He made a significant contribution to our knowledge of the fossils of the Hunsrück Slate, and of the lower Devonian of Germany in general, through his collecting and research (Opitz 1930, 1931, 1932, 1935). His work provided an important platform for subsequent investigations. Opitz's book was sold out shortly after publication, no subsequent edition was

Figure 1 Illustrations from Ferdinand Roemer's (1862) paper describing new asteroids and crinoids from Bundenbach which included the first published illustrations of Hunsrück Slate fossils: *Urasterella* (top) and *Euzonosoma* (bottom).

published, and copies are now extremely rare. Opitz (1890–1940) was one of a number of eminent German paleontologists, including F. Broili (1874–1946), R. Richter (1881–1957), and W.M. Lehmann (1880–1959), and amateurs who made the Hunsrück Slate a central focus of their work. Following the death of Lehmann, who was still active in research in the late 1950s, interest in this Fossil-Lagerstätte waned. It was widely believed that there was little prospect of making important new discoveries in the Hunsrück Slate, or of increasing our knowledge of the Lower Devonian. In addition, there was a crisis in the slate-mining industry during the late 1950s, and by 1963 nearly all mining had ceased, especially in the most important region paleontologically, around the village of Bundenbach. Thus, from that time on, new specimens rarely came to light.

Only a small group of professional earth scientists and committed amateurs continued to investigate the Hunsrück Slate and its fossils, and they intensified their efforts from 1969 on. Fritz Kutscher of the Geological Survey of Hessen, and Wilhelm Stürmer, a chemical-physicist and director of a group developing radiological methods at the Siemens Corporation, instigated new research. Kutscher had published his thesis on the geology of the Hunsrück Slate in the early 1930s (Kutscher 1931). Stürmer had been one of Richter's students and, combining his skills as radiologist and paleontologist, he developed efficient methods of investigating fossils using x-rays (Stürmer 1968, 1969*a,b*, 1970*a,b*, 1984, Stürmer and Bergström 1973). The Kaisergrube mine near Gemünden (Fig. 2) was leased for scientific excavation and research under Stürmer's direction. His radiographs of Hunsrück Slate fossils (e.g. Figs. 127, 130) became internationally known through exhibitions and two folders, each with eight images, published by the Siemens Corporation. He brought together a group of international specialists to study the results of his radiological investigations, and between 1970 and 1989 systematic papers on several groups of Hunsrück Slate fossils were published (e.g. Stürmer 1985, Stürmer and Bergström 1973, 1976, 1978, Bergström *et al.* 1980, 1987, 1989, Stanley and Stürmer 1983, 1987, Fauchald *et al.* 1986, 1988).

Günther Brassel and Christoph Bartels enjoyed a close and long-standing collaboration in their research on the Hunsrück Slate with Fritz Kutscher and Wilhelm Stürmer. Stürmer died in 1986, Kutscher two years later, and we acknowledge our grateful debt to these outstanding paleontologists and geologists, and close friends. Many of the discoveries and interpretations presented in this book are theirs, some of them the result of our cooperation and collaboration. Without their involvement this book would not have been written. Derek Briggs was privileged to be part of a group assembled by Stürmer in Erlangen in

Figure 2 The Kaisergrube mine near Gemünden, site of paleontological excavations and research under the direction of W. Stürmer from 1969 to 1985.

1984 to investigate *Nahecaris* and other phyllocarids. He subsequently investigated the pyritization of the Hunsrück Slate fossils in collaboration with Rob Raiswell and his colleagues at the University of Leeds. It was this research that initiated his collaboration with Bartels and Brassel and led to the preparation of this revised and updated English edition of their book (Bartels and Brassel 1990).

The first part of this book deals with general aspects of the Hunsrück Slate, its importance as a source of roof slates for many centuries (Bartels 1986), its geological history, environment of deposition, and the preservation and diagenesis of the fossils (Chapters 1 to 3). The unique evidence provided by the Hunsrück Slate fossils for the nature of marine life during the early Devonian is the subject of the second part. The best known of the Hunsrück Slate fossils are the arthropods, echinoderms and vertebrates, which have made the Bundenbach region famous the world over. Other elements of the fauna also provide unique evidence of the history of soft-bodied animals, including ctenophores and poly-chaete worms. The remarkably preserved trace fossils are also receiving attention. The most important elements of the Hunsrück Slate biota are described in out-line in Chapters 4 to 9. We have attempted to provide high quality photographs of representative examples. Inevitably only a selection of taxa is included; it is neither feasible nor desirable to describe every species in detail. For convenience we have used the classification adopted in Benton (1993) *The Fossil Record 2* except where otherwise indicated. Where the affinity of groups is controversial alternative interpretations are rehearsed. The third part of the book (Chapters 10 and 11) discusses the collection and preparation of the fossils and the applica-tion of different techniques to the Hunsrück Slate, and outlines possible areas of future research. The German book (Bartels and Brassel 1990) upon which this one is based was written to publicize the importance of the Hunsrück Slate and its fossils within Germany. We hope that, by reaching a wider audience, this new English version will encourage further research and lead to increased interest in the fascinating roof slates of the Rheinisches Schiefergebirge.

The authors are grateful for photographs supplied by G. Brassel jun., W.v. Koenigswald (Bonn). A. Opel (Bochum), A. Nestler (Bad Kreuznach), C. Underwood (Bristol), S. Powell (Bristol), radiographs from W. Blind and I. Steubing (Giessen), and diagrams from U. Dittmar, O. Sutcliffe, and M. Wuttke. Figures from the literature were redrawn by D. Woelfel and P. Orr. Many experts kindly read and provided critical comments on parts of the text: Uwe Dittmar (Mayen) and Owen Sutcliffe (Bristol) – geological setting, trace fossils, Rob Raiswell (Leeds) – pyritization, Dianne Edwards (Cardiff) – plants, Michael Wuttke (Mainz) – sponges, Alan Thomas (Birmingham) – trilobites, Michael Simms (Cheltenham) and William Ausich (Columbus, Ohio) – crinoids, Andy Gale (London) and Dan Blake (Champaign, Illinois) – starfishes, Chris Paul (Liverpool) – homalozoans and other echinoderms, and Peter Forey (London) – fishes. The Museum Idar-Oberstein and its director, Alfred Peth, kindly granted permission to publish this revised and expanded English version. Many individuals have helped over the years with hints, advice and discussion of our work on the Hunsrück Slate and helped us bring this book to fruition. We are particularly grateful for all the friendly assistance we have been afforded in Bundenbach and in other areas of roof-slate production by the slate miners, and the roof-slate producing companies and their representatives. Briggs's research on the Hunsrück Slate was funded by NERC GR9/518. The collaboration between Briggs and Bartels was supported by a British German Academic Research Collaboration Programme funded by the British Council and the German Academic Exchange Service (DAAD).

The sources of illustrations are as follows: C. Bartels (Deutsches Bergbau-Museum): 3, 5, 11, 12, 17, 18, 29, 52, 68, 82, 86, 89, 103, 105, 118, 153, 154, 157, 165, 168, 189, 190, 204, 209, 221, 222, 224, 225, 235; W. Blind: cover, 32, 37, 53A,B, 88, 94, 100, 174A, 178, 187A,B, 197, 236; W. v. Koenigswald: 126; G. Brassel. jun.: 48, 54, 83, 87, 96, 109, 114, 118, 146, 169, 196, 220, 233, 234; C. Underwood: 36; Landesbildstelle Rheinland-Pfalz: 2; Landeshauptarchiv Koblenz: 6; A. Opel (Deutsches Bergbau Museum): 1, 20, 21, 23, 25–28, 31, 33–35, 38–40, 42, 45–47, 49, 56–62, 66, 67, 69, 72–75, 78, 80, 82, 84, 85, 90–92, 95, 97, 99, 102, 106–108, 111–113, 115, 116, 120, 121, 128, 131, 135–142, 144, 145, 147–152, 155, 156, 158–164, 166, 167, 170–173, 174B–177, 179–184, 186, 188, 191–194, 198–201, 203, 205–208, 211, 213, 217, 223, 226–230, 232, 237; A. Nestler: 195; W. Stürmer: 19, 22, 30, 41, 43, 50A,B, 63, 65, 70, 71A,B, 77, 79, 93, 110, 117, 123, 127, 130, 143, 185, 202; O. Sutcliffe: 4, 16, 24; M. Wuttke: 51; D. Woelfel (Deutsches Bergbau-Museum): 55, 64, 98, 101, 104, 122, 124, 125, 129, 132, 210; P. Orr: 14, 43, 51, 76, 124, 127, 231.

Specimen and radiograph repositories are abbreviated in the figure legends as follows: HS, Bartels collection, Deutsches Bergbau Museum, Bochum; SNG, Brassel Collection, Naturmuseum Senckenberg, Frankfurt am Main; WS, W. Stürmer radiograph (Naturmuseum Senckenberg); WB, W. Blind radiograph (Institut für Angewandte Geowissenschaften, Giessen).

Christoph Bartels, Deutsches Bergbau-Museum, Am Bergbaumuseum 28, D-44791, Bochum, Germany

Derek E.G. Briggs, Department of Geology, University of Bristol, Wills Memorial Building, Queen's Road, Bristol BS8 1RJ, UK

Günther Brassel, Osterallee 75, D-24944 Flensburg-Mürwik, Germany

Part I
The Hunsrück Slate

1 Mining the Hunsrück Slate

The mining areas

The river Rhine and its tributaries incised deep valleys into the uplands of the Rhenish Massif dividing them into the blocks of Hunsrück and Taunus, Eifel and Westerwald. The topography of undulating elevated plains and steep valleys is largely lithologically controlled. The higher mountains are formed by quarzites, while the softer shales make up the lower ground of the plains. Slate-covered roofs and walls are a typical element of the architecture of towns and villages both along the river Rhine and in the uplands east and west of the river (Bartels 1986). Indeed the German name, Rheinisches *Schiefer*gebirge (i.e. Rhenish *slate* mountains), stresses the importance of shales and particularly roof slates in the underlying geology.

The valleys of the rivers Rhine (Fig. 3) and Mosel in particular were sites of ancient human settlement (Zschocke 1970, Bartels 1986) due to the favourable climate and their importance as early north–south (Rhine) and east–west (Mosel, Lahn) routes. Several types of stone and minerals were exploited in this region from prehistoric times. Millstone (Mühlstein) lava, for example, was extracted around Mayen in the south-eastern Eifel mountains. Millstones were exported from this area during the first millennium BC to sites as far away as east of the Ural mountains (Röder 1970, Hambloch 1913). Roof slate is one of the most commercially important lithologies in the region. Where great thicknesses of uniform muds were deposited with only rare thin sands and silts, this resulted in extensive developments of slates that could be mined and quarried for roofing and for facing walls.

Roof slates are widespread in the middle Hunsrück area. Here the valleys of two small rivers, the Hahnenbach and the Simmerbach, are deeply incised into

Figure 3 The Rhine valley near Kaub. The view looking north-westwards shows the famous medieval castle Rheinpfalz, which served as a customs post, built on an island where the Hunsrück Slate crops out in the middle of the river. The Rhein roof-slate mine lies on the west bank and the Wilhelms Erbstollen mine, which produced roof slate until 1972, on the east bank.

the plateau. The slopes of these river valleys (and those of their tributary streams) allow access to the slate deposits. Most of the famous fossils have been found here, around Bundenbach and Gemünden (Fig. 4).

There is a second region of roof slate mining in the Rhine valley around the small towns of Kaub on the right bank and Bacharach on the left bank of the river. The Rhine and its small tributaries have exposed the nearly vertically oriented roof slate layers. This region of slate production continues from the Rhine valley for a considerable distance into the Taunus mountains along the valley of the Wisper river, and as far east as the town of Bad Schwalbach. There are several smaller areas of roof slate mining to the south of the river Mosel – in

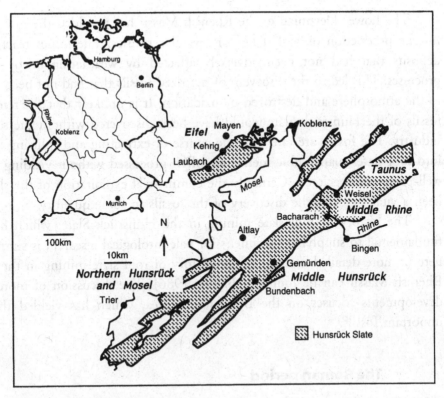

Figure 4 The Hunsrück Slate region.

the valley of the Ruwer near Trier, south-west of Wintrich, around Zell, and in the valleys of other tributaries of the Mosel near Altlay, Hahn and Treis-Karden.

The most important commercial slate production developed in the area around Mayen in the south-western Eifel mountains. Here mining extended from Müllenbach and Laubach in the west to Ochtendung east of Mayen (Bartels 1986). These slate deposits lay close to the margin of the Middle Rhine Depression (Mittel-Rheinische Senke) where settlements grew up at the intersection of old trade routes along the Rhine and the valleys of the Mosel and Lahn. Here a combination of fertile soils and important deposits of iron, copper, lead and zinc ores as well as building materials (Hambloch 1913, Hashagen 1913, Oberbergamt Rheinland-Pfalz 1963), resulted in a high population density and consequent demand for roof slate. For a brief period slate mining was important around Isenburg in the south-west of the Westerwald mountains in the valley of the river Sayn. There was also a long tradition of slate production in the Wissenbach Slate (Wissenbacher Schiefer) in the Lahn valley around Dietz but these deposits are younger than the Hunsrück Slate (they are latest Emsian to Eifelian in age) and are not considered here (see Völker 1978).

The Lower Devonian of the Rhenish Massif has provided the raw material for the production of roof slates for several centuries. The mines reached slate deposits that had not been adversely affected by weathering or by tectonic processes. This led to the discovery of pyritized fossils that had not been exposed to the atmosphere and destroyed by oxidation. It is unlikely that the remarkable fossils of the Hunsrück Slate would have been discovered without the roof slate industry. The fossils are rare – in the course of extracting and splitting the rock into roof slates many cubic metres are often processed without yielding a single well-preserved specimen. Certainly the commercial exploitation of the slates has been a vital factor in the discovery of the fossils in some quantity.

The history of roof slate mining in the Hunsrück Slate, which has been fundamental to supplying specimens for paleontological research, is summarized here (a more detailed account of the history of roof slate mining in the eastern Rhenish Massif can be found in Bartels 1986). The discussion of more recent developments focuses on the Bundenbach area, which has yielded the most important fossils.

The Roman period

The earliest evidence of the use of roof slates in western Germany dates from the time of the Roman Empire. It is not known, however, whether the Romans introduced the use of roof slates (just as they introduced the cultivation of grapes for wine production) or whether they simply continued an ancient tradition in the region along the river Rhine and its tributaries. What is clear is that slates were in widespread use in this area a short time after the arrival of the Romans (Fig. 5). The towers and fortifications of the Roman town of Xanten on the Rhine in north-western Germany, for example, were roofed with slate. The techniques used to work the slate and make the roofs are still in use today (Freckmann and Wierschem 1982). Remains of slate roofs are common at Roman sites all over western Germany. In his poem *Mosella*, written in 368, the Roman poet Ausonius described 'towering roofs', presumably made of slate, in the villages and towns along the Mosel. The sandstone roof of a famous Roman tomb in Igel near Trier imitates the style of a slate roof (Fischer 1970, Quiring 1931, 1932). We know almost nothing about the Roman slate mines, however, although sites where they quarried other building materials are documented. The slates used in Xanten are very similar to those from the Mayen region. There is some circumstantial evidence that the Romans mined slate in the middle Rhine valley and near Bacharach, but proof is lacking.

Figure 5 The ruins of a Roman craftsman's house outside the walls of the Niederbieber Castle near Koblenz, which was destroyed in AD 259. Some slabs of slate that were used in the roof have accumulated in the area indicated.

The Middle Ages and post-medieval time

The history of roof slate mining from the end of the Roman period to the beginning of the fourteenth century is completely unknown. Unfortunately the provenance of the materials used in medieval buildings has attracted little interest, in spite of its potential as a significant source of information on the economy of the time. All we know is that castles and other medieval buildings were roofed with slate.

The oldest known document relating to mining in the Hunsrück Slate authenticates the sale of a share of a roof slate mine near Bacharach on 13 October 1300. A document of 3 January 1355 proves that slate was being mined near Kaub on the opposite bank of the Rhine (Tinnefeld 1989). The financial records of several town governments in Germany during the Middle Ages list payments to slaters for the upkeep of civic buildings, testifying to the importance of slate roofs. In 1363, for example, there were 24 slaters and only one thatcher working in Trier, and similar records from several other towns (Kremer 1951) emphasize the importance of slate for roofing at least in urban areas. Slate mining must have been relatively widespread in the region. Extensive documentation of medieval slate mining survives in the archives of the town of Goslar in lower Saxony (Burkhart 1938), but this is outside the region under consideration here.

In 1605 Friedrich Hellbach, a chronicler of the Hunsrück region, reported that slate mining was so extensive in the region of the Middle Rhine valley that slate was sent 'all over the countryside' (Frölich 1924). This is confirmed by the records for the same year of the town of Gross-Umstadt in Hassia, some 120 miles south-east of the slate-producing region, which state that slate for the roof of the town hall was bought in Bacharach (Tinnefeld 1989).

Numerous important public buildings, both ecclesiastical and secular, were erected during the seventeenth and eighteenth centuries as testimony to the might and splendour of the Age of Absolutism. Huge quantities of slate were used in these buildings (Bartels 1986). The 1767 map of 'Schollmuther Leyenberge' (Fig. 6), for example, is evidence of extensive slate mining at this time around the village of Irmenach in the Hunsrück region. Slate mining was also important in eastern France during the eighteenth century. Copperplate engravings dating from about 1750 show the mining and processing of slabs (Fougeroux de Bondaroy 1763). The techniques illustrated (Fig. 7) are the same as those used in the Rhenish Massif (Bartels 1986).

It is not known whether slate mining and manufacturing in early times

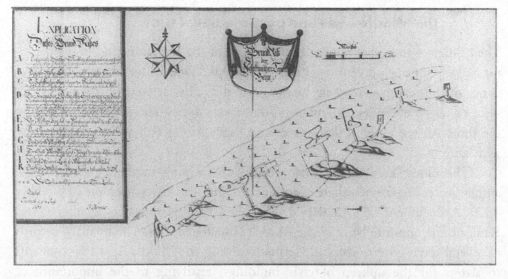

Figure 6 A map of the roof-slate mines near Irmenach in the Hunsrück area in 1767. Each of the mines consists of a short addit leading to a single excavation. (Mine surveyor J. Thomas. Photograph courtesy of the State Archive, Koblenz (33/9577).)

Figure 7 Roof-slate production during the 18th century. The tools illustrated here are similar to those still in use in the Hunsrück area today (Fougeroux de Bondaroy, 1763, *Die Kunst den Schiefer aus den Steinbrüchen zu brechen*, Berlin, Pl. IV.)

resulted in any knowledge or understanding of fossils. No information on collecting activities prior to the mid nineteenth century has survived. Fossils from the Hunsrück Slate are unknown in scholar's collections, or in cabinets of curiosities in great houses. Presumably only slate workers were aware of the 'Figuren' (figures) in the stone, the term used by the local people of the time to refer to fossils. Perhaps they were considered to be sports of nature, as was the tendency in post-medieval times.

Industrialization

The onset of industrialization during the last third of the eighteenth century resulted in a rapid expansion of the roof slate industry. Slate was mainly produced by farmers during the winter as a sideline to their farming activities. Extraction from surface workings, at least in significant quantities, was generally impossible because of the deep narrow valleys, combined with the steep orientation of many of the roof slate deposits (Bartels 1986). Thus from the outset slate was mined underground rather than quarried. Contemporary accounts allow us to build up a comprehensive picture of the operation of a slate mine in the Eifel area (typical of mines in the Rhenish Massif) during the first half of the nineteenth century.

A road, along which the slate was hauled, led from the mine. This road had been cut into the rock with difficulty, and with the passage of time the wheels of the heavy carriages had made deep grooves in the surface. The road ended at the mine dump, the surface of which was roughly level. Here the slabs of roof slate were standing upright in long rows, waiting to be transported away. There was an open hut on the top of the dump, essentially a temporary roof to provide some protection for the splitters who sat on the ground working the blocks of slate. On the side of the mountain a low mound marked where the gallery opened. On entering the gallery with a small, smoking oil-lamp called a 'frog' the miners had to bend to avoid the ceiling. After some metres the low passageway ended in a sheer drop. A contemporary account (dated 1819: see Bartels 1986) records that: 'a crude spiral staircase, built from discarded pieces of slate, descended 50 feet vertically to the actual working, which extended down the surface of the slate bed for over 3 Lachter [1 Lachter is roughly 2.20 m] to groundlevel, and along the slate level for 4 Lachter [roughly 9 m]. The working area was protected on both sides by residual rock pillars.' The visitors had then reached the deepest point of the irregular working which descended steeply into the mountain along the bedding plane of the rock. The excavation here reached a maximum height of about 20 m, and the small oil lamps could not illuminate its full extent. Some miners were working below at a 'wall', several metres high and long. Parts of the excavation had already been filled with rubble. Walls built of discarded pieces of slate prevented this material from collapsing. The workers had constructed a platform some metres high on a framework of tree trunks to enable them to reach the higher parts of the slate 'wall' that was currently being worked. As blocks of slate were excavated they were slid down the steeply inclined surface of the slate layer. This was a dangerous business, because of the risk of blocks hitting the tree trunks which held up the working platform. The workers split the blocks into smaller pieces on the floor of the excavation. The same contemporary account noted that 'It will be obvious that this is an arduous way to mine slate, as the workers have to carry the blocks of raw roof slate on their backs from the workings up the winding stairway to the surface [see Fig. 8]. I considered myself fortunate to have safely negotiated the staircase without that weight on my back.' The account also observed that 'these roof slate mines are all very similar, rather like the burrows of moles or foxes, with no regular pattern; individual workings within each mine are all different. The workers do not operate any system but work up or down, always pursuing the best slate for as long as possible, until a rock fall puts an end to this groping around; then they dig a new entrance and the process starts again in just the same way' (see Bartels 1986, pp. 46–8).

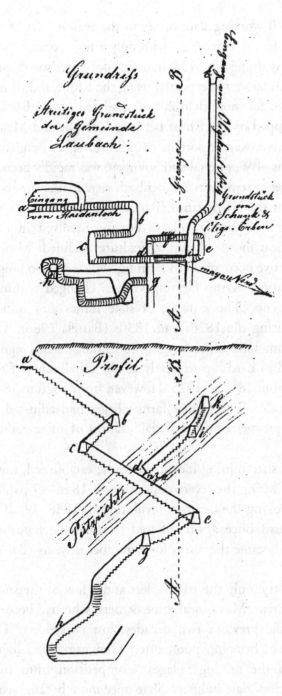

Figure 8 Plan and vertical section of the Heidenloch roof-slate mine at Laubach in the south-eastern Eifel in 1858 showing the galleries and staircases for the transport of raw slate blocks. (State Archive, Düsseldorf, Bergamt Düren, no. 121, p. 191.)

By 1810 there were some 40 working slate mines in the region west of the Rhine, most of them small-scale family enterprises with only a few workers. Slate mining was still expanding, and within a few years trade in roof slates was booming. A significant part of the production was exported using the Rhine and Mosel. Around 1820 roof slate accounted for more than 20% (calculated on the basis of loading weights) of all goods shipped on the Rhine between Cologne and Mainz. Regions east and west of the Rhine were important markets for slate. Bergisches Land, for example, with the towns of Wuppertal and Solingen, was rapidly becoming industrialized. By 1830 villages as remote as Bundenbach were marketing slates over long distances, even as far as the Netherlands (Bartels 1985*a*, 1994*b*).

More and more people in the mining villages became involved in roof slate production as the population increased and agriculture declined. Mining rapidly developed into an alternative occupation to farming, which was no longer adequate to support the population. Many farming villages changed within a short time into mining settlements. The number of slate mines and miners increased particularly rapidly during the 1820s and 1830s (Bartels 1986). The change in the local economy from subsistence farming to mining caused significant social changes and resulted in hardship to nearly all the inhabitants of the region, particularly in the years from 1817 to 1825. However, from 1826 to 1845 the situation was much more stable. The former farm villages had adjusted to the new economy. Trade firms prospered and a social stratum of mineworkers had developed.

Almost as soon as the roof slate mining industry became established, however, it suffered a rapid collapse during the severe depression of 1846–49 (which was an important factor in provoking the German revolution of 1848–49). The inhabitants of the Rhenish Massif once again suffered considerable hardship. Famine, disease, and lawlessness became the norm for a number of years (Bartels 1986).

Within the roof slate industry, only the wholesalers and a few of the small mining companies survived the crisis. Many local mine owners, who had become comparatively wealthy during the previous two decades, lost everything. The boom in slate production ceased, bringing poor, often even miserable, living standards to the inhabitants of the mining villages. Competition from rival materials, especially tiles, hit the roof slate industry. Slate once more became what it had been for centuries, an expensive luxury normally used only for the roofs of public buildings (Bartels 1985*a,b*, 1986).

A new boom in slate production was only initiated following the German–French war of 1870–71 and the revival of the German Reich. The new

German nationalism was often expressed in buildings, and it became fashionable to use roof slate in an 'old German' manner. This, together with a general economic revival, led to a growth of slate mining. This period was characterized by the rapid development of technology and science, and intense industrialization (Bartels 1986). Now the fossils of the Hunsrück Slate attracted interest for the first time. Geologists and miners, as well as private fossil collectors, began to collect the 'Figuren' (figures or specimens) in the slate and to publish their finds (Kutscher 1969a).

The increasing demand for slate favoured mainly the bigger companies that could raise the capital necessary for the development of more extensive mines. A major factor in the growth of these mines was the advent of a new transport system, the railway. The Middle Rhine region, with mines around Bacharach and Kaub and in the vicinity of Mayen and the Mosel valley, was linked to the railway network during the 1870s and 1880s, facilitating further expansion of mining in this area. This led to the development of several large modern slate mines, the largest in the region employing about 150 workers. This was still small, however, compared with mines where the geological setting was more favourable. In north Wales, eastern France (around Angers) and south-eastern Germany (Thuringia), for example, workforces of many hundreds (sometimes over a thousand) were employed.

The twentieth century

Up to 1880 slate-mining technology had not changed significantly except that small mines with short addits and only one working chamber were replaced by galleries some 100 m in length leading to a couple of workings. Thus the rather random excavations of the early years with their miserable economic conditions had been replaced by a more systematic approach (Bartels 1986). During the last two decades of the nineteenth century deeper shafts were sunk in the larger slate mines. Steam engines were used for haulage. Pneumatic hammers were introduced around 1910 and slate mining became partly mechanized. The mine installations and buildings on the surface were subdivided into smaller plants, which were interconnected and joined to the shaft entrance by railways (Figs. 9, 10).

Within a short time many of the small slate mines were closed down as they could not compete with the bigger companies. Bundenbach provides a good example. The village was not connected to the railway network. Plans to open a bigger mine were abandoned around 1925 and the slate industry persisted in

Figure 9 The Müllenbacher Dachschieferwerk roof-slate mine in the Kaulenbach Valley near Müllenbach, south-eastern Eifel, around 1910. (Peters collection, Müllenbach.)

Figure 10 The Katzenberg roof-slate mine at Mayen in 1910 as illustrated on the company letter-head.

small rather primitive mines, similar to those of a century earlier (Bartels 1985*a,b*, 1994*b*; Fischer 1959*a,b*). Whereas there had been some 90 to 100 miners working in Bundenbach at the turn of the century, by 1910 their number had dwindled considerably and 40 to 50 men from Bundenbach were working in the mines of the Rhine valley. Here they earned considerably higher wages than those paid in the small mines nearer home. They had to walk for four to five hours at the beginning and end of each week to and from the little town of Bacharach on the Rhine. The management of the 'Rhine' roof slate mine there built a lodging house to accommodate the miners from out of town during the week.

During the years of inflation after 1922 and the world economic crisis of 1929–1930 many bigger mining companies cut back or even stopped production. Ironically this led to an increase in slate production around Bundenbach. In this remote region people accepted wages, working hours and conditions that would not have been tolerated in more developed areas. In contrast to workers in the towns, they were also engaged in farming and forestry which guaranteed a supply of food and fuel. Most of the families owned a small house and some land where they grew vegetables and fruits, and many kept a pig or at least some chickens and a goose. With their basic needs assured, they were prepared to accept hard working conditions in the small mines to supplement their income. So, as slate production declined elsewhere, the small Bundenbach mines, most of them family enterprises, enjoyed something of a boom from about 1929. Some of the small firms employed between 30 and 50 workers, their success a result not least of the exceptional quality of the roof slate of the Bundenbach area.

In the years before the second world war there was a general boom in the building industry in Germany, part of it in the construction of military barracks. Slate roofs were regarded as a particularly German feature so, as in the first world war, the military was a major consumer of slate. Shortly after the war began, however, slate mining went into decline as many miners had to join the army. Production almost ceased following 1943. The devastation of war resulted in a new boom after 1948. During the reconstruction period and at the beginning of the German 'economic miracle' ('Wirtschaftswunder') which followed, the high demand for slate allowed smaller mines to flourish as well the larger companies that had dominated the industry since the early 1920s. From 1958 on, however, many of these small concerns, and some bigger slate mines, closed down. They could not compete with synthetic roof slates which could be produced cheaply due to the very low oil prices of the late 1950s and the 1960s. Synthetic 'slates' soon dominated the market, particularly for use on new buildings.

The German 'economic miracle' created a great deal of new employment and provided relatively well paid jobs, particularly for younger men, with much more attractive conditions than mining. This prompted many slate miners to leave the industry (and the mining villages) for new types of work. Slate production stopped in several areas, including that around Bundenbach. By 1970 only two mines near Mayen, one in the Taunus mountains east of the Rhine and two in the Hunsrück area, continued to work. Alongside the extraction and production of local material, raw slate was imported, particularly from Spain and Portugal, and manufactured into roof slates in the traditional mining areas. One firm in Bundenbach continued to operate using only imported slate.

The international oil crisis of 1973 resulted in a rapid increase in the price of synthetic 'slates' as well as prompting considerable efforts to conserve energy by insulating buildings against heat loss. The good insulating qualities of natural slate have led to an increase in demand in the last 20 years. This has not, however, been accompanied by a new boom in slate mining in the Rhenish Massif. Existing companies have enjoyed a new prosperity, but mainly by importing larger quantities of slate, particularly from Spain, where production costs are considerably lower than in Germany. The rise in demand has also meant a new lease of life for a small firm run by just two workers in Bundenbach. This enterprise had survived for years under a permanent threat of bankruptcy by working the remaining slate in numerous abandoned mines. Now they were able to reopen the Eschenbach–Bocksberg roof slate mine. The excellent quality of the slate allowed

Figure 11 The Eschenbach–Bocksberg quarry at Bundenbach.

them to expand their operation which has now been run as a quarry for about 20 years and employs some 30 workers (Fig. 11). Most significantly for paleontology the slate layers in Eschenbach–Bocksberg quarry yield the famous pyritized 'Bundenbach fossils' in some quantity.

The Eschenbach–Bocksberg mine is the only one (of more than 50) that has been reopened in the area around Bundenbach where the most spectacular fossils are found; it is very unlikely that any others will be worked again in the future. Elsewhere two pits near Mayen, Katzenberg and Margaretha, continue to produce roof slate (Fig. 12). The Rhine mine near Bacharach only produces

Figure 12 Methods used in the production of roof-slate today. A. Splitting the slate by hand; B. Slate saw; C. Splitting with pneumatic tools; D. Mechanized shaping.

slate powder, which is used in the chemical industry and in the manufacture of ceramics. A slate quarry was opened some time ago near the village of Altlay by a firm from the Bundenbach area, but it lies outside the region of productive fossil beds. The Kreuzberg mine near Weisel in the Taunus mountains did yield some interesting fossils but, in spite of being profitable, it was closed in 1982 when the two tenants retired.

In the light of the history of slate mining since the second world war, any further expansion of roof slate mining in Germany is unlikely. Imported slate from Spain remains cheaper, and small enterprises like the Eschenbach–Bocksberg quarry are particularly vulnerable to crises in the building industry. It is difficult to predict how much longer the unique exposures of roof slates near Bundenbach will remain accessible. In spite of their scientific importance they have been largely neglected until recently by geological institutes and museums in Germany. Only now are their sedimentology, fossilization and diagenesis receiving the attention they merit (see pp. 266–8). It would indeed be ironic if this remarkable fossil occurrence were lost to paleontologists before it is fully researched.

2 Geological setting

The age of the Hunsrück Slate

The term Hunsrückschiefer (Hunsrück Slate) has been rather loosely applied to a sequence of Lower Devonian sedimentary rocks that crops out in the Rhenish Massif and is dominated by extensive deposits of mudstones which have been metamorphosed to slates. In the past, as now, the Hunsrück Slate has been regarded as a sedimentary facies or depositional setting (Kutscher 1931, Richter 1931, Solle 1950), although some authors have attempted to treat it as a stratigraphic unit (Solle 1950, Mittmeyer 1974, 1980*a*, 1982). The latter approach is misleading, however, as conditions for slate deposition migrated in time and space from the late Gedinnian until the end of the Middle Devonian over a considerable area of the Rhenish Massif. In general the muddy facies was first established in the north-west, and migrated through time to the south-east. Deposition was in a number of basins separated by more upstanding structures (swells) and correlation across the outcrop is difficult.

Although the Hunsrück Slate has no formal stratigraphic status it can be regarded as roughly equivalent to a lithostratigraphic 'Group' (Mittmeyer 1974, 1980*a*). German stratigraphers divide the Lower Devonian into Gedinnian, Siegenian and Emsian (the international stage names are listed on Fig. 13). The Hunsrück Slate was deposited between the end of the Siegenian and the middle of the Lower Emsian (*ca* 392 to 388 million years ago: Harland *et al.* 1990) (Fig. 13). The famous pyritized fossils are restricted in space and time to a small part of this sequence. The most spectacularly preserved examples come from the area around Bundenbach and Gemünden. The Hunsrück Slate corresponds to three local subdivisions or Unterstufen: the Herdorf (Upper Siegenian), Ulmen and Singhofen (Lower Emsian), named after the type localities. Mittmeyer (1980*a*)

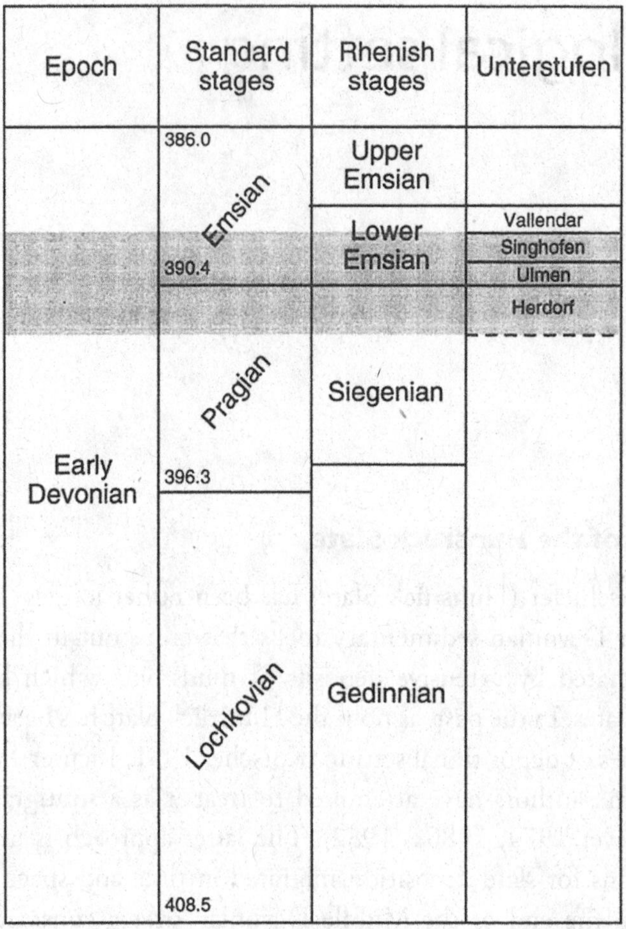

Epoch	Standard stages	Rhenish stages	Unterstufen
	386.0	Upper Emsian	
	Emsian	Lower Emsian	Vallendar
			Singhofen
	390.4		Ulmen
			Herdorf
	Pragian	Siegenian	
Early Devonian	396.3		
	Lochkovian	Gedinnian	
	408.5		

Figure 13 Stratigraphic position of the Hunsrück Slate (ages, in millions of years; from Harland *et al.* 1990).

argued for a more restrictive definition, corresponding to the Ulmen-Unterstufe only. This would exclude examples of the typical roof slate facies that occur in beds of different ages outside the classic Hunsrück area around Bundenbach and Gemünden. Mittmeyer (1980*a*) identified the Ulmen-Unterstufe on the basis of the absence of tuff horizons which characterize the other subdivisions of the sequence. The stratigraphy is also subdivided on the basis of brachiopods (Mittmeyer 1982). Bartels, however, cast doubt on the basis for identifying and correlating the Ulmen-Unterstufe by discovering tuff in the Hunsrück Slate near Bundenbach (Bartels and Kneidl 1981, Kirnbauer and Wendorf 1995). The Ulmen subdivision, which has not been universally accepted since it was established by Solle in 1950, may even be of equivalent age to the Singhofen subdivision (as argued by Nöring 1939). Much work on the stratigraphy of the Hunsrück Slate remains to be done. Modern treatments are attempting to subdivide the Hunsrück Slate on the basis of lithostratigraphic markers and micro-fossils (Alberti 1982*a,b*, 1983, Sutcliffe 1997*b*). Many areas are largely unknown and even the age of the classic area is not well constrained.

Plate tectonic setting

Seismic investigations show that the southern margin of the Rhenish Massif is bounded by a major dislocation, the South Hunsrück and Taunus Borderzone (Anderle 1987) that descends several kilometres into the crust. This separates the Devonian and older rocks of the Massif to the north-west from Upper Carboniferous/Lower Permian rocks (Rotliegendes) in the Saar–Nahe Basin to the southeast. Thus there is no direct evidence for the nature of the Devonian to the south. The Borderzone forms part of a major tectonic lineament in the crust of Western Europe which has been active since the early Devonian. It cor-responds in position to a major collision zone between two continental plates (Ahorner and Murawski 1975).

The details of the plate tectonic setting of the Hunsrück area are contro-versial (Franke 1995). Interpretations based on paleomagnetic data (Perroud *et al.* 1984, Van der Voo 1988) and models based on tectonics (e.g Anderle 1987, Franke 1989) have not been fully integrated. The Hunsrück area was associated with a microplate or terrane known as Armorica (Perroud *et al.* 1984, Van der Voo 1988). During the Silurian and early Devonian Armorica was attached to the northern margin of the great southern continent of Gondwana. It had been picked up as Gondwana drifted northward (from about 70° S into the tropics at 10°–20° S to collide with Laurentia. This collision led to the closure of Iapetus and the Medio-European Ocean by the late Silurian–early Devonian and resulted in the Caledonian orogeny.

Early plate-tectonic interpretations implied that the Hunsrück Slate sedi-ments were deposited on the shores of Baltica which was still separated from Armorica to the south by the Medio-European Ocean (Burrett 1972). However, the timing of the docking of Armorica with the northern continents (Laurentia and Baltica) is not well constrained (Perroud *et al.* 1984, Van der Voo 1988). The Carboniferous dates for the deformation of the Hunsrück Slate (Ahrendt *et al.* 1983, Massonne 1995) suggest that the sequence was deposited after Armorica collided with the northern continents. This is supported by the position of the Hunsrück Slate adjacent to the South Hunsrück and Taunus Borderzone which defines the southern margin of Armorica where it collided with Gondwana during the later Variscan orogeny.

The nature of the Hunsrück depositional basin is unknown; it may have been a response to back-arc extension or to thinning and subsidence over an extension of the former spreading axis of Iapetus (Franke 1989, Berthelsen 1992, Franke 1995). This gave rise to the short-lived Rheno-Hercynian or Lizard–Giessen–Harz Ocean separating the Mid-German Crystalline High (Mitteldeutsche Schwelle) to the south from the Soonwald Swell to the north. The Soonwald Swell formed the southern margin of the Central Hünsruck Basin, the area in which the sediments of the Middle Hunsrück, Middle Rhine and Wisper regions were deposited. The depositional context has been elucidated by Dittmar (1996) (Fig. 14). To the north the basin was bounded by a second positive structure, the Central Hunsrück Swell, which became emergent during the Emsian. Thus the Hunsrück Slate sediment in the classic Bundenbach–Gemünden area was derived from both the north and south (Herrgesell 1978). Further north the Hunsrück Slate of the Northern Hunsrück and Eifel regions was deposited in the Mosel–Hunsrück Basin.

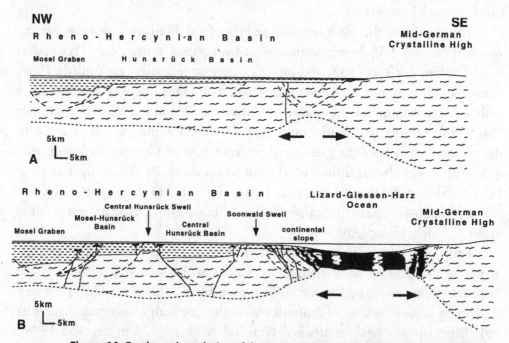

Figure 14 Geodynamic evolution of the south-western Rhenish Massif and adjacent areas: (A) late Siegenian, showing the accumulation of sediments on continental crust in the Hunsrück Basin and Moselle Graben; (B) the transition between early and late Emsian (Unterems to Oberems) showing the spreading Lizard–Giessen–Harz Ocean and the continued accumulation of sediments, including Hunsrück Slate, in the basins to the northwest of the Soonwald Swell (after Dittmar 1996).

It has been suggested that, following the deposition of the Hunsrück Slate, Gondwana returned southward during the Devonian with the formation of an ocean separating Armorica, now attached to the Old Red Sandstone Continent, from Africa. The evidence for this Proto-Tethys Ocean, however, is equivocal (see Franke *et al.* 1995) (Fig. 15). What is clear, however, is that closure of the Rheno–Hercynian Ocean was completed by Carboniferous times as Gondwana moved northward and collided with Armorica and the Old Red Sandstone Continent. It is this continent-continent collision (comparable to the India–Asia collision that uplifted the Himalayas) that formed the huge continent Pangaea and led to the Variscan or Hercynian Orogeny (Van der Voo 1988). This orogeny affected an area including the Rhenish Massif, the Harz Mountains, the Pyrenees, Brittany and south-west Britain, and the Appalachians in North America.

The Variscan Orogeny left the Hunsrück sediments in a flat-lying stack of imbricate thrust-sheets (Anderle 1987), and subjected them to low grade metamorphism which produced the slaty cleavage. The organic maturity of spores indicates that the Hunsrück Slate reached temperatures of about 400 °C in the course of the orogeny (Wolf 1978, Ecke *et al.* 1985). This metamorphic event gives an age of about 325 million years, i.e. late Early Carboniferous (Ahrendt *et al.* 1983, Massonne 1995). The collision zone corresponds to the South Hunsrück and Taunus Borderzone. The area of deposition of the original Hunsrück sediments was much more extensive than that covered by the Hunsrück Slate today. Folding and thrusting led to substantial shortening. This reached more than 50% at the south-eastern margin of the Hunsrück Slate (in the Phyllite Zone) and 25–30% in the Bundenbach–Gemünden area. In some cases, where the fossils are inclined to the cleavage, they can be used as strain indicators (e.g. Fig. 37).

Events after the Devonian

Towards the end of the early Emsian the deposition of thick sequences of mud that gave rise to the Hunsrück Slate was largely replaced by the coarser sediments of the Normal Facies which were deposited in shallower water around the margins of the Old Red Sandstone Continent. Uplift led to the formation of a land mass on the Central Hunsrück Swell, the Hunsrück Island of Solle (1970) (see Fig. 14). Extensive tidal flats developed on the eastern and northern flanks of this large island (Solle 1970). Its rapid emergence is consistent with the evidence for shallow water sedimentation in the Hunsrück Slate lithologies.

Between the Emsian and the Lower Carboniferous most of the region of

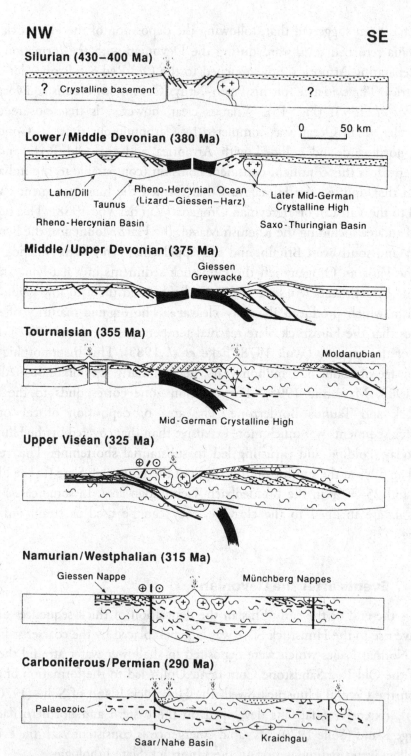

NW

SE

Silurian (430–400 Ma)

? Crystalline basement   ?

Lower / Middle Devonian (380 Ma)

0   50 km

Lahn/Dill   Rheno-Hercynian Ocean   Later Mid-German
      Taunus   (Lizard–Giessen–Harz)   Crystalline High

Rheno-Hercynian Basin   Saxo-Thuringian Basin

Middle / Upper Devonian (375 Ma)

Giessen
Greywacke

Tournaisian (355 Ma)

Moldanubian

Mid-German Crystalline High

Upper Viséan (325 Ma)

Namurian / Westphalian (315 Ma)

Giessen Nappe   Münchberg Nappes

Carboniferous / Permian (290 Ma)

Palaeozoic

Saar/Nahe Basin   Kraichgau

Figure 15 Plate tectonic scenario for the Rheno-Hercynian and Saxothuringian belt (from Franke and Oncken 1990).

the Rhenish Massif was covered by the sea and some 1000 m of marine sediments were deposited (Anderle 1987). These sediments were also deformed by the Variscan Orogeny. North-west to south-east compression produced the cleavage that allows the rock to be split into roofing slates. In the axes of folds, the bedding and cleavage are at a high angle and fossils are difficult to discover. On the fold limbs, however, the cleavage and bedding become parallel and the rock can be split along or very close to the bedding to reveal the fossils. This situation occurs most frequently in the vicinity of Bundenbach and Gemünden as well as in the southern Taunus region. On the southern margin of the Hunsrück area more intense deformation resulted in refolded phyllites (Anderle 1987, Massonne 1995).

In places the tectonic events resulted in local extension. The injection of hydrothermal fluids along the fracture planes gave rise to deposits of copper, lead, zinc and silver. These deposits provided the basis for the former metal mining industry in this region. Copper of very high quality was mined from the Hosenberg mine near Fischbach about 20 km south-west of Bundenbach during the sixteenth and seventeenth centuries. Some lead, copper and zinc was also mined at Friedrichsfeld mine near Bundenbach (Wild 1983).

During the late stages of the Variscan Orogeny (Anderle 1987, Franke *et al.* 1995) uplift and erosion in the Hunsrück area was associated with downwarping and subsidence of the Saar–Nahe Basin to the south. This basin formed in the synclinal part of a large fold on the southern margin of the Rhenish Massif. The northern margin of the syncline consisted of steep to vertically dipping thrusts. This trough accumulated 7000 m to 10 000 m of sediment during the Carboniferous and Permian, the Rotliegendes ('red beds'), derived initially from the north, subsequently from the south. Volcanic rocks within this sequence formed a thick covering in part of the Nahe–Pfalz region. These volcanic rocks are the source of the famous agates, amethyst and other minerals which have been mined for many years around Idar-Oberstein in a region justly renowned among mineralogists.

Uplift of the Rhenish Massif from Late Jurassic to Eocene led to erosion of the Variscan mountains, including the area of Hunsrück Slate. During the Oligocene the peneplain produced was flooded by a brief marine transgression evidenced by the preservation of sediments yielding foraminiferans. Only a few areas of high relief formed small islands. Uplift from the Miocene to the present is associated with the evolution of the Rhine Ruhr graben, including the Middle Rhine Depression (Neuwieder Becken). This uplift was accompanied by volcanic activity in the south-eastern Eifel and the formation of volcanic lakes (e.g. the Laacher See).

The Tertiary and Quaternary elevation of the peneplain resulted in the reju-venation of the drainage system. The Rhine extended its catchment further south. The topography of the Rhenish Massif today is dominated by the old peneplain that resulted from the erosion of the Variscan uplands, and the younger valleys which are eroding down toward the level of the Rhine graben floor.

Paleogeography and sedimentary environment

The Hunsrück Slate sea was formerly interpreted as a narrow seaway between the Old Red Sandstone Continent and the Mid-German Crystalline High to the south. The Hunsrück Slate was thought to have been deposited in a channel-like depression, connected to both east and west with the open sea. Solle (1950) com-pared it to a muddy area off the modern North Sea coast protected by a major offshore sand bank. He regarded the Taunus Quartzite as representing the sand bank.

In modern interpretations the major element in the paleogeography of the Hunsrück area during the Lower Devonian remains the Old Red Sandstone Continent, a product of earlier Caledonian uplift, which lay to the north. The coastline ran north-east to south-west fluctuating in position along a line from Essen through Aachen to Namur. The Old Red Sandstone Continent is named after the red sediments that formed under a warm climate in semi-arid condi-tions. Terrestrial facies accumulated on the continent while marine sediments were deposited to the south. Thus in northern Europe (in parts of northern Britain, for example) Old Red Sandstone facies of the same age as the Hunsrück Slate are found. Equally, marine facies, different in character to the Hunsrück Slate, were deposited at the same time elsewhere (e.g. slates, sandstones and limestones in southern Britain).

Erosion of the Old Red Sandstone Continent supplied river systems which transported mud and sand to the south (see Fig. 14). Sediment was deposited in the classic Bundenbach–Gemünden area in a north-east to south-west trending offshore basin termed the Central Hunsrück Basin (Dittmar 1996; equivalent to the Wisper Trog of Mittmeyer 1980a). Downwarping allowed the Hunsrück Slate to reach thicknesses that Mittmeyer considered to have been as much as 3000 m (1980) or even 4600 m (1982). Recent work indicates that these values may be an overestimate. Dittmar (1996) calculated thicknesses of 2300 m on the south-eastern margin of the Central Hunsrück Basin and 3750 m on the north-western margin. The sequence with roof slates ('Kauber Schichten') was estimated as 950 m between Bundenbach and Gemünden (Dittmar 1996). Our own observations suggest that even this may be an exaggeration due to tectonic repetition that is very difficult to detect in such fine-grained lithologies.

Sediment built out into the Central Hunsrück Basin in lobes or fans. Gravel was deposited nearer the coast. Finer sediment was carried in suspension by turbidity currents and transported offshore to areas of quiet water where it gave rise to muddy bottom conditions. Evidence of paleocurrent directions is difficult to obtain in the Hunsrück Slate. Around Gemünden, however, current indicators are mainly oriented north-east to south-west (Herrgesell 1978). These indicate that deposition on the distal part of the fans was affected by currents deflected parallel to the axis of the basin. Some current directions indicate a sediment source to the south (Herrgesell 1978) suggesting that the basin shallowed not only to the north-west but also to an island lying to the south-east (as envisaged by Solle 1950). This is borne out by the position of the Soonwald Swell to the south of the Central Hunsrück Basin (Dittmar 1996). Mud accumulated in depressions between the prograding submarine fans. In some shelf areas water depths were very shallow and conditions were occasionally emergent. Herrgesell (1978) argued that shallow water, perhaps brackish, conditions prevailed in some Hunsrück Slate horizons around Gemünden, especially in layers yielding abundant specimens of the agnathan *Drepanaspis gemündensis* and lacking fully marine forms such as echinoderms. Shallow depths are indicated by the absence of grading in the sandy layers, and the presence of bipolar cross-bedding, lag deposits, and coquinas. Richter (1931, 1954) also recorded marks from Gemünden that he interpreted as the product of foam on the sediment surface. This suggests that here, at least, the Hunsrück Slate was deposited in shallow water, occasionally even tidally influenced, and interfingered with the Taunus Quartzite which accumulated in ridges and depressions. It is generally accepted that water depths represented by the Hunsrück Slate did not exceed 200 m. The upper part of the water column was fully oxygenated, and there is no evidence that dysaerobic conditions were established for long periods of time even in deeper depressions.

Research on the detailed sedimentology of the Hunsrück Slate (Sutcliffe 1997b) emphasizes how variable it is lithologically even in the area around Bundenbach and Gemünden. Polished slabs and thin-sections reveal frequent changes in sedimentation. Background sedimentation from suspension was relatively insignificant. Silty and sandy laminations commonly occur within the clays. Beds of sand a few centimetres to several tens of centimetres thick, many showing grading, punctuate the sequence (Fig. 16). These sandy lenses are not extensive laterally, however, perhaps reflecting submarine channelling. Stürmer and Bergström (1973) calculated that the *average* rate of sedimentation was

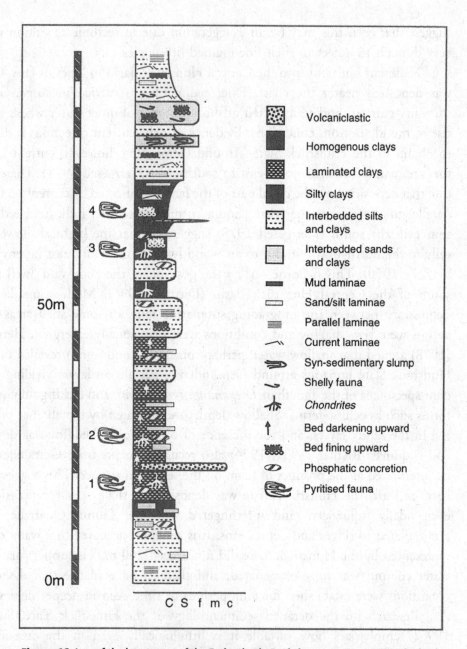

	Volcaniclastic
	Homogenous clays
	Laminated clays
	Silty clays
	Interbedded silts and clays
	Interbedded sands and clays
—	Mud laminae
▭	Sand/silt laminae
▤	Parallel laminae
⌒	Current laminae
⤳	Syn-sedimentary slump
~	Shelly fauna
⚘	*Chondrites*
↑	Bed darkening upward
∞	Bed fining upward
⊘	Phosphatic concretion
⥤	Pyritised fauna

Figure 16 Log of the lower part of the Eschenbach–Bocksberg quarry near Bundenbach as an illustration of the sedimentological features of the Hunsrück Slate.

2 mm of mud per year (prior to dewatering and compaction). However, as they recognized, much of the sedimentation was episodic, the result of turbidity currents generated perhaps in response to tropical storms. Sediment was reworked and even eroded (Herrgesell 1978) leading to changes in the conditions on the sea floor where the fauna was living (Fig. 16).

The Lahn–Dill region and the adjacent Sauerland formed a centre of volcanic activity to the north that may have contributed volcaniclastic sediment to the Hunsrück Slate sequence (Franke 1995). A 30–100 cm thick layer of tuff in the Bundenbach area is called 'Hans' by the roof slate miners, supposedly after the worker who discovered that it can be used as a marker bed to determine the position of the roof slates within the sequence. 'Hans' can be followed for about 8 km south-west to north-east in the Bundenbach area. This tuff was deposited as a composite channel fill and thus it varies in thickness, thinning towards channel margins. Spilitic lava has also been reported in the Hunsrück Slate of Bundenbach (Kirnbauer 1986). Tectonic movements associated with the volcanic activity may have caused the slumping or contorted bedding which is evident, for example, in the Eschenbach–Bocksberg quarry. Volcanic eruptions presumably also initiated sediment transport by turbidity currents (Kirnbauer and Wendorf 1995).

3   Paleoecology and preservation

The Hunsrück Slate as a Konservat-Lagerstätte

The concept of Fossil-Lagerstätten was introduced by Seilacher in 1970 to refer to exceptional fossil deposits. Two categories were identified, Konzentrat-Lagerstätten where fossils are preserved in great abundance, and Konservat-Lagerstätten where it is the quality of preservation that is remarkable. Konservat-Lagerstätten range from examples of complete articulated skeletons (of echinoderms or vertebrates, for example) to settings where traces of soft tissues (i.e. those that are not biomineralized in life) survive. Soft-tissue preservations have received more attention than other Fossil-Lagerstätten (see reviews in Briggs 1991, Allison and Briggs 1991a,b). Few Konservat-Lagerstätten are confined to a restricted area on a single bedding plane. Many represent multiple events, albeit it in the same setting. Most are interspersed with levels where the preservation is much less spectacular. It is not surprising, therefore, that Konservat-Lagerstätten are often referred to by the name of the larger unit in which they occur – units that vary in the amount of time and space that they represent. The soft-bodied fossils of the Middle Cambrian Burgess Shale, for example, come from some of a large number of individual burial events; the remarkable fossils of the Solnhofen Limestone are scattered over an area of hundreds of square kilometres. The Hunsrück Slate sequence includes silts and sands interbedded with the roof slates. The remarkable pyritized fossils are confined to a few horizons, often less than 1 cm thick, and generally of restricted lateral extent. These horizons occur within a thickness of some 1000 m and are confined to a small area within more than 400 km^2 of total outcrop. Thus the levels containing the critical fossils are atypical and it is misleading to apply the term Konservat-Lagerstätte to the Hunsrück Slate as a whole.

More than 600 mines have exploited the different areas of Hunsrück Slate (Bartels 1986). Thus the roof slate lithology has been intensively scrutinized for years as it was split into slates about 5 mm in thickness. Nevertheless horizons that merit the term Konservat-Lagerstätte have only been found in the Bundenbach and Gemünden area, and the quality of preservation varies even within these. Complete skeletons of delicate organisms like crinoids, starfishes, and complex arthropods are common, but soft parts are rarely preserved. The most spectacular fossils have been found at mines near Breitenthal, Bundenbach and Gemünden. Similar examples have been found at a few horizons in the Wisper valley in the Taunus region (Figs. 17, 41, 148–150) and in the Middle Rhine valley around Bacharach and Kaub, but the preservation is inferior to that in the Bundenbach area. Comparably preserved fossils are otherwise extremely rare (Fig. 18). In areas where the cleavage is at a high angle to the bedding, however, exceptionally preserved fossils may have gone unnoticed because they are usually destroyed during the processing of the slate.

The collections in museums and other scientific institutions give a false impression of the normal state of fossil preservation in the Hunsrück Slate, because only the most spectacular specimens are acquired. Previous interpretations of the Hunsrück Slate have focused only on these rare exceptionally preserved

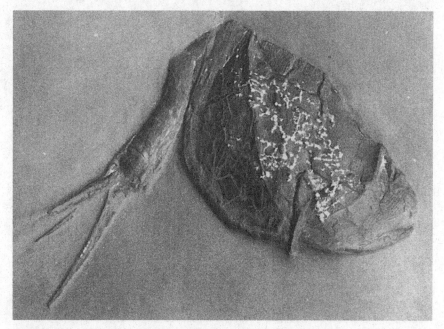

Figure 17 A specimen of the crustacean *Nahecaris* sp. The tissues themselves are not pyritized but numerous euhedral crystals of late pyrite have formed on the surface of the carapace, Kreuzberg mine, Weisel, Taunus (×0.9; HS 213).

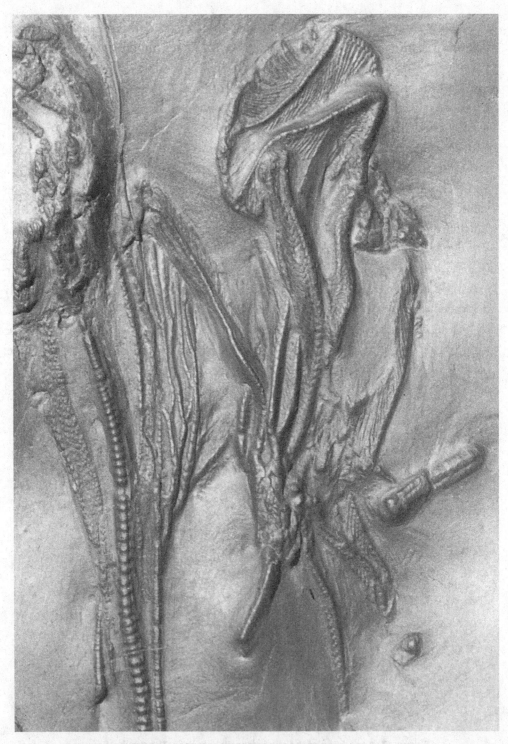

Figure 18 *Acanthocrinus* sp. and undetermined crinoids which have not been pyritized, Margaretha mine, Mayen (× 1.6; HS 107).

fossils, neglecting the bulk of the paleontological evidence. Observations during slate processing at the Eschenbach–Bocksberg quarry near Bundenbach, where the best preserved fossils occur, reveal that most horizons yield only fragments of organisms in different states of decay and disarticulation (Figs. 19–21). Where these are abundant the slate is unsuitable for roofing. Very often tonnes of slate are worked without finding a single well-preserved fossil, and where they do occur they tend to be confined to a single horizon of limited lateral extent.

Horizons yielding exceptionally preserved fossils occur in nearly all the roof slate mines around Bundenbach, but there is no regularity in their distribution. The same taxa occur elsewhere in the Hunsrück Slate, but usually preserved as fragments. The preservation at some levels in the Rhenish Normal Facies is similar to that of the Hunsrück Slate, except that traces of the soft parts do not survive.

The fine-grained muddy nature of the sediment alone is not sufficient to explain the extraordinary preservation in the Bundenbach area. Pyritized soft parts are not preserved in most of the Hunsrück Slate, even in the quarries and mines around the type locality at Bundenbach. Complex fossils such as crinoids, starfish and other echinoderms have disarticulated. More robust skeletal elements such as the dermal plates of fishes, the skeleton of corals, and shells of

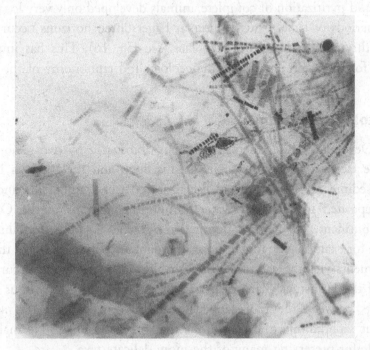

Figure 19 Disarticulated crinoids, Kaisergrube mine, Hunsrück (× 1.0; radiograph WS 3198).

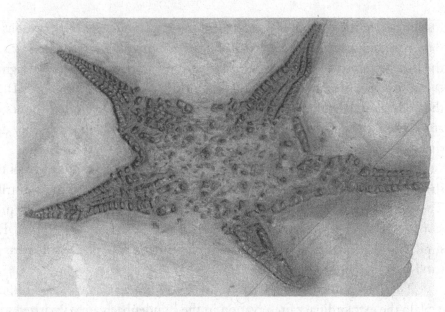

Figure 20 A specimen of *Euzonosoma* sp. with the plates of the body disc largely disarticulated, perhaps as a result of the activity of scavengers. Eschenbach–Bocksberg mine, Bundenbach (×0.8; HS 198).

cephalopods, are more common (see Figs. 66–84). Clearly conditions for intact burial and rapid pyritization of complete animals developed only very locally and for limited periods of time. Thus Konservat-Lagerstätten horizons occur *within* rather than throughout the Hunsrück Slate (see Fig. 16). This has important implications for the interpretation of the ecology and taphonomy of the fauna.

Ecological setting

The Lower Devonian of the Rhenish Massif is dominated by two major facies, the roof slate facies which is the focus of this book and the Rhenish Normal Facies. The sediments of the Normal Facies are coarser grained silts and sands that were deposited in shallower water around the margins of the Old Red Sandstone Continent. The fauna is dominated by shelly organisms; molluscs and brachiopods, for example, are much more abundant and diverse than in the slate. Sandier sediments yielding a typical Rhenish Normal Facies fauna occur in the south-east of the Hunsrück Slate area near Wisper valley in the Taunus region. The contrasts between the roof-slate facies and Normal Facies are not only ecological but taphonomic; the conditions that prevailed in the Normal Facies are unsuitable for preserving many of the more delicate taxa.

The ecological setting of the roof slate facies of the Hunsrück Slate has

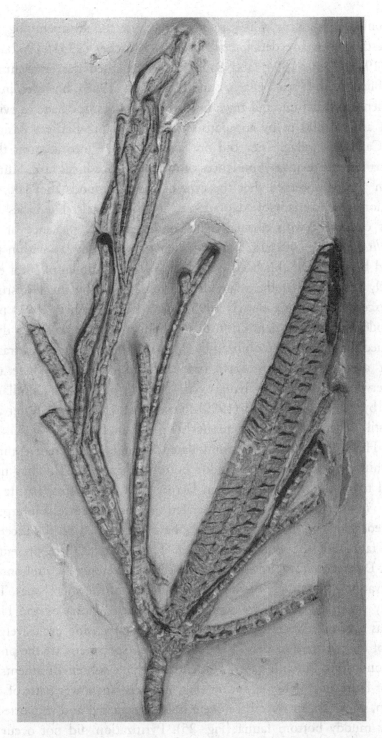

Figure 21 A specimen of *Gastrocrinus giganteus* W.E. Schmidt, 1934, which lost the stem and much of the arms prior to burial, presumably as a result of disarticulation during transport, Eschenbach–Bocksberg mine, Bundenbach (× 1.0; HS 94).

proved controversial. This is at least partly because the sedimentology had not been investigated in any detail until recently. Kutscher (1931, 1962a, 1974a), Richter (1931, 1936, 1954) and Solle (1950) regarded the environment as a tidally influenced shelf sea. Seilacher and Hemleben (1966), however, interpreted the sedimentary structures and trace fossils of the Hunsrück Slate as evidence of depths of around 800 m by comparison with the Santa Barbara Basin off the coast of California today. They did not, however, take into account the entire ichnofauna nor the relative abundances of different trace fossil taxa. Stürmer and Bergström (1973) observed that the eyes of the arthropods show none of the modifications characteristic of adaptation to low light levels. The fishes, likewise, have eyes consistent with normal light conditions. The presence of the alga *Receptaculites* (see Fig. 40) also indicates that the sea floor lay within the zone penetrated by light. On this basis, together with the sedimentological evidence, it is widely accepted that the water depths represented by the Hunsrück Slate did not exceed 200 m (see Bartels 1995). Brett and Seilacher (1991) presented a new model relating the deposition of the Hunsrück Slate to storm dynamics. They argued that sediment mobilized by storms in shallow water was transported to deeper areas below storm wave base as mud turbidites. These turbidity events overwhelmed the biota living on the bottom in a manner similar to that envisaged by Bartels and Brassel (1990). Nevertheless the deeper water interpretation is still adhered to in some quarters (e.g. Erben 1994).

The Hunsrück Slate sea accommodated a range of different communities living within, on and above the muddy bottom (Figs. 22, 23). They no doubt responded to subtle local variations in factors such as currents, nature of sedimentation, temperature, light, and associated organisms. Previous interpretations of the paleoecology of the Hunsrück Slate have been based on the exceptionally preserved faunas of the Bundenbach area. The same taxa are present in the Konservat-Lagerstätten as in the sequence; the difference is taphonomic, in the preservation of complete organisms with details of the soft tissues. Thus the remarkably preserved fossils are representative of the fauna of the Hunsrück Slate sea as a whole. We must beware, however, of focusing exclusively on the evidence of the pyritized fossils. By definition these specimens are the product of unusual conditions. Sutcliffe (1997b) identified three 'sub-environments' within the area of Hunsrück Slate deposition (Fig. 24). The shallower parts of the submarine fan, above storm wave base, were fully oxygenated and supported a normal shelly muddy bottom fauna (Fig. 25). Pyritization did not occur in this sub-environment. The minor depositional lobes, where channel distributaries reached the lower part of the fan, are characterized by coarsening-upward

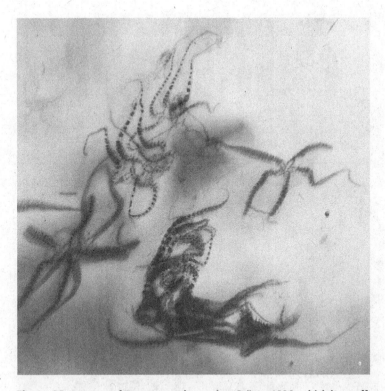

Figure 22 A group of *Furcaster palaeozoicus* Stürtz, 1886, which has suffered obrution (rapid burial). The variation in the density of the image in different parts of the radiograph reflects differences in the degree to which the fossils have been pyritized, Eschenbach–Bocksberg mine, Bundenbach (×0.65; HS 146; radiograph WS 12341).

sequences. Here the sediment deposited by density currents retained sufficient oxygen to support an infauna. These burrowing animals destroyed any more delicate potential body fossils, although the burrows themselves (similar to *Planolites*) became sites of pyritization. Finally the classic pyritized fossils of Bundenbach and Gemünden are preserved in interchannel areas. Here the water was oxygenated allowing muddy bottom communities to become established. Occasional density currents, however, buried the animals, the thickness of sediment protecting them from scavengers. The sediment rapidly became anoxic leading to conditions that promoted pyritization.

The crinoids were one of the most important groups inhabiting muddy sea floors during the early Devonian (Roemer 1863, Follman 1887, Jaekel 1895, Schmidt 1934, 1941, Bartels and Brassel 1990) (see Fig. 18). They colonized the bottom as isolated individuals, in clusters, or as large concentrations or 'meadows'. They had a free-swimming larval stage, but became attached to the sea bed as adults. In the Hunsrück Slate they are often attached to fragments of

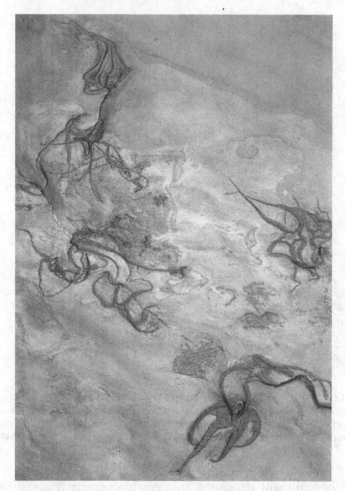

Figure 23 A large group of animals overwhelmed by a rapid influx of sediment: ten *Furcaster palaeozoicus* Stürtz, 1886, a *Taeniaster beneckei* (Stürtz 1886) a small crinoid and five partly fragmented specimens of the arthropod *Mimetaster hexagonalis* (Gürich, 1931), Eschenbach–Bocksberg mine, Bundenbach (×0.4; HS 538).

shells, including those of large orthocone cephalopods. Many crinoids developed root-like extensions of the stem (cirri) which anchored and supported them in the muddy sediment (Fig. 26). A number of specimens preserve these structures at a lower level in the sediment than the calyx and arms. It is likely that at least the ends of the cirri were partially covered by sediment during the life of the crinoid. Perhaps they even grew into the sediment. Slabs displaying several crinoids often include examples of different stages of development from the tiniest juveniles to large adults (Figs. 27, 172). Juvenile crinoids, some only a few millimetres long and yet to develop arms, are frequently attached to adult specimens (see Figs. 26, 35, 157, 172). The preservation of many of these crinoid specimens

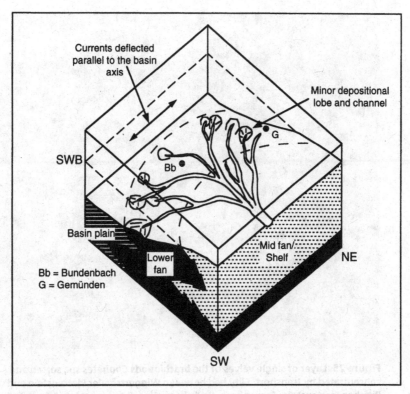

Currents deflected
parallel to the basin
axis

Minor depositional
lobe and channel

G

SWB

Bb

Basin plain

Mid fan/
Shelf

NE

Lower
fan

Bb = Bundenbach
G = Gemünden

SW

Figure 24 Restoration of the depositional environment of the Hunsrück Slate
(SWB = storm wave base).

provides unequivocal evidence that they were buried in life position (see Figs.
157, 162, 172). Although others have been uprooted, they have probably been
transported only a very short distance. Even the earliest papers on the Hunsrück
Slate concluded that the crinoids were buried where they lived on the sea bed
(Jaekel 1895).

It is clear, therefore, that the Hunsrück Slate sea floor was a living
environment. This is borne out by evidence of the activities of arthropods. The
trilobite *Chotecops* is abundant and occurs over almost the entire area of Hunsrück
Slate outcrop. The occurrence of numerous trilobite trackways on the sediment
surface indicates that the trilobites were alive. The traces were not obliterated by
erosion or bioturbation of the sediment. Such trackways are even found on the
top of turbidite beds showing that life continued immediately after an obrution
event. Preservation of these and other traces indicate that the animals lived in
quiet conditions on the sea floor.

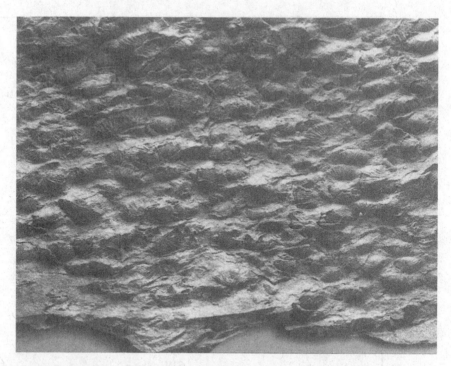

Figure 25 Layer of single valves of the brachiopods *Chonetes* sp., sorted and concentrated by transport. Silty bed between Wingertsheller Plattenstein and Pittchen Plattenstein, Eschenbach–Bocksberg mine, Bundenbach (×0.6; HS 542).

Community structure

Field observations of the Hunsrück Slate demonstrate that, while there is an overall dominance of fine-grained sediment, there is considerable variability from locality to locality and within the sequence. Superimposed on these variations in sediment type are differences in the preservation (including pyritization) and content of both body and trace fossils. There is considerable variation in the fossil content from layer to layer at individual localities within the Bundenbach area. The Hunsrück Slate has not been studied in sufficient detail to allow the pattern of these changes in time and space to be elucidated.

Communities dominated by crinoids, asteroids and ophiuroids, together with trilobites and other arthropods, existed in close proximity in oxygenated, current-influenced conditions within the photic zone. Due to the rarity of the fossils and the lack of systematic sampling (almost all collecting relies on splitting during commercial slate production) no statistical analysis of the Hunsrück Slate fauna has been carried out. Some general observations, however, can be offered. The great majority of organisms preserved in the Hunsrück Slate were benthic. This no doubt reflects not only the muddy bottom environment

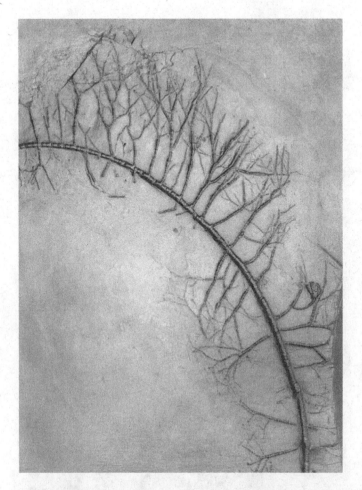

Figure 26 The stem of the crinoid *Dicirrocrinus comtus* W.E. Schmidt, 1934, showing the network of branching cirri which supported the animal and prevented it from sinking into the sediment. Note the attached juvenile of *Codiacrinus* on the right-hand side. Eschenbach–Bocksberg mine, Bundenbach (× 0.7; HS 104).

but also the mode of deposition. Animals living up in the water column are much less likely to be overwhelmed by clouds of suspended sediment carried by turbidity currents.

The sessile benthos was dominated by crinoids, which were the most common organisms in the Hunsrück Slate sea. They were anchored to the sea floor by coiling or other modifications of the distal part of the stem, or as a result of attachment to shelly fragments such as bivalves, cephalopods, or trilobites (see Figs. 35, 82, 156, 162, 165). Only one crinoid, the inadunate *Senariocrinus*, may have been free living (see Fig. 159). Like modern crinoids the Hunsrück Slate forms fed on organic particles and small organisms filtered from the water column. Other members of the sessile benthos include the sponges, corals,

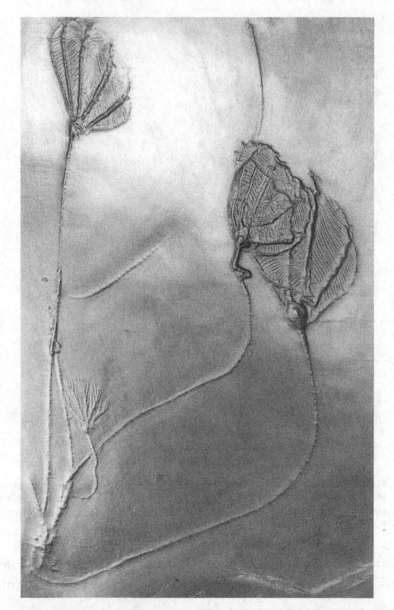

Figure 27 Three adults and a juvenile *Hapalocrinus frechi* Jaekel, 1895, buried in life position. Eschenbach–Bocksberg mine, Bundenbach (×0.8; private collection).

conulariids, some bivalves, brachiopods, and bryozoans. The benthos also included algae, the most striking of which are the very rare receptaculitids (see Fig. 40).

The most important members of the vagrant benthos were the arthropods, starfishes (asteroids and ophiuroids), and fishes. Most, if not all, of the Hunsrück

Slate arthropods lived on the sea bottom, walking on and swimming near the sediment surface. Some may have burrowed in the uppermost layers of the sediment. Their activities are reflected in a diversity of locomotion traces (see Figs. 221–224). *Mimetaster* and *Vachonisia* were probably deposit feeders. Many of the other arthropods, however, were predators or at least scavengers. In most cases the evidence is indirect; identifiable gut contents or examples of wounded prey are very rare. *Nahecaris* had robust mandibles which must have been capable of breaking even biomineralized shells. The appendages of *Cheloniellon*, the chelicerate *Weinbergina*, and the trilobites *Asteropyge* and *Chotecops* had proximal spines that formed gnathobases suitable for capturing and cutting up prey, probably mainly soft-bodied worms. There is more direct evidence that the giant pycnogonid *Palaeoisopus* (Figs. 28, 130–132) was a predator. It could walk or swim using its long legs, and was armed with large pincers or chelifores. Specimens normally occur in horizons that are rich in other animals, including crinoids, suggesting an abundant food source. *Palaeoisopus* may have preyed on crinoid meadows (Bergström *et al.* 1980), and its activities may account for the occasional specimens of crinoids and starfishes that are found with severed arms.

The starfishes (asteroids and ophiuroids) (Fig. 29) were slower moving members of the vagrant benthos. Unlike the crinoids, however, their mode of life may not have been the same as that of their modern counterparts. The large size of the mouth, and lack of evidence for a predatory mode of life, suggests that they

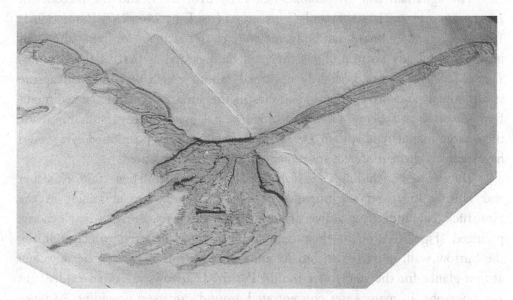

Figure 28 Large complete specimen of the pycnogonid *Palaeoisopus problematicus* Broili, 1928, Eschenbach–Bocksberg mine, Bundenbach (×0.4; HS 582).

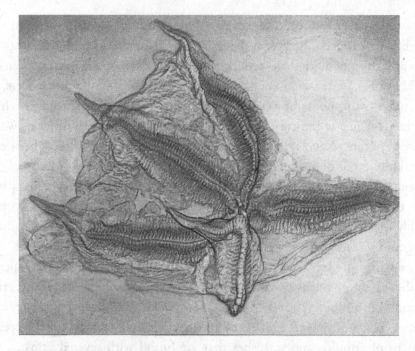

Figure 29 A strongly pyritized example of *Loriolaster mirabilis* Stürtz, 1886, Eschenbach–Bocksberg mine, Bundenbach (×0.6; HS 374).

were deposit feeders. The Hunsrück Slate gastropods presumably also fed by ingesting sediment.

The agnathan fish *Drepanaspis* (see Figs. 210, 211), and the placoderms *Lunaspis* (see Figs. 214, 215) and *Gemuendina* (see Figs. 216, 217) (Gross 1961, 1963 *a,b*), all had a broad, flat body and were clearly adapted for living and feeding on the muddy bottom. Thus they can be regarded as part of the benthos. *Drepanaspis* had a broad, slightly upturned mouth which was used to capture small animals and other organic particles in the surface layers of the sediment. Both *Lunaspis* and *Gemuendina* had broad lateral fins, which presumably made them more manoeuvrable than *Drepanaspis*. Their dorsally positioned eyes would have enabled them to detect movement in the overlying water.

Trace fossils, including *Chondrites* (Seilacher and Hemleben 1966, Kutscher and Horn 1963) and the burrows of deposit feeders, are abundant in the Hunsrück Slate indicating a diversity of infaunal elements. Burrows may become pyritized (Fig. 30) due to their content of organic material (mucous lining the burrow wall, or faeces within it) and simple forms may even be mistaken at first glance for the stems of crinoids! Pyritized burrows are often evident on x-radiographs, in many cases concentrated around carcasses, providing evidence of the activity of scavengers (see Fig. 227). Many of these traces were doubtless

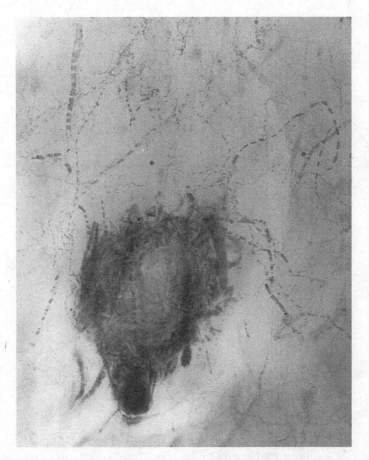

Figure 30 An accumulation of dacryoconarids and small orthocone shells which may represent a coprolite or a burrow fill. The radiograph shows pyritized traces of numerous burrows of worm-like organisms, some of which appear to have been partly filled by strings of faeces. Kaisergrube mine, Gemünden (× 1.2; WS 533).

made by worms, including polychaetes which are also occasionally represented by pyritized body fossils.

The soft parts of actively swimming organisms are rarely preserved because they decay following death in the water column. Shells of nautiloid cephalopods, however, of various sizes, are found at virtually all Hunsrück Slate outcrops suggesting that these swimming predators occurred in relatively large numbers (Figs. 31, 71, 72). The largest predators were the arthrodires. These fishes are known only from incomplete specimens, but fragments of the spines and armour of *Tityosteus* indicate that individuals up to two metres long swam in the Hunsrück Slate sea. The more passive chondrophorines and comb jellies were carried by water currents nearer the surface, capturing small organisms in their tentacles as they do today.

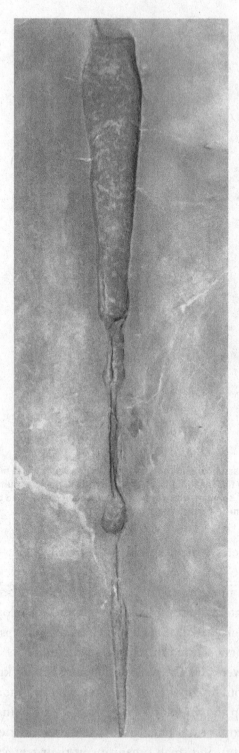

Figure 31 Non-pyritized orthocone cephalopod, Mariaschacht mine, Laienkaul near Laubach in the Eifel region (×0.6; HS 42).

Death and burial

The death of an organism is normally followed by the decay of the soft tissues including muscles, skin, connective tissue, and even skeletal tissues like the cuticle of arthropods where these are not biomineralized. Materials are broken down and recycled through the activity of microorganisms. The hard parts (e.g. shells, bones, teeth, spines) are more durable, and therefore more likely to become fossilized.

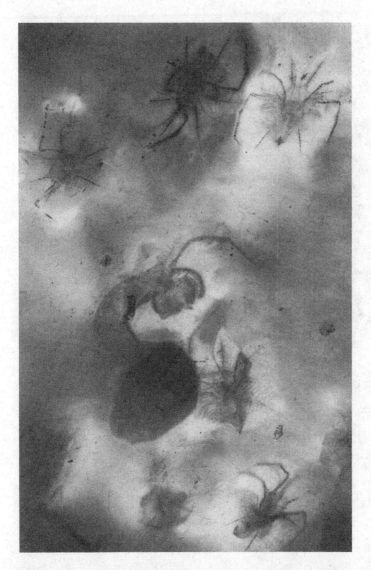

Figure 32 Part of a large slab with more than 20 individuals of *Mimetaster hexagonalis* (Gürich, 1931), buried in a variety of orientations to bedding as a result of turbulent transport. Eschenbach–Bocksberg mine, Bundenbach (cf. Whittington 1985, p. 34, Fig. 3.8) (×0.75; Natural History Museum, Mainz, PWL 1993/265-LS, radiograph WB6).

They are usually composed of calcite or aragonite (calcium carbonate) or apatite (calcium phosphate). Skeletons normally disarticulate and the elements scatter once the soft tissues that hold them together have decayed. Only rarely are they buried rapidly enough to remain intact. Thus the majority of Hunsrück Slate localities yield disarticulated skeletons, and no soft parts are preserved. This style of preservation is essentially similar to that in the Rhenish Normal Facies.

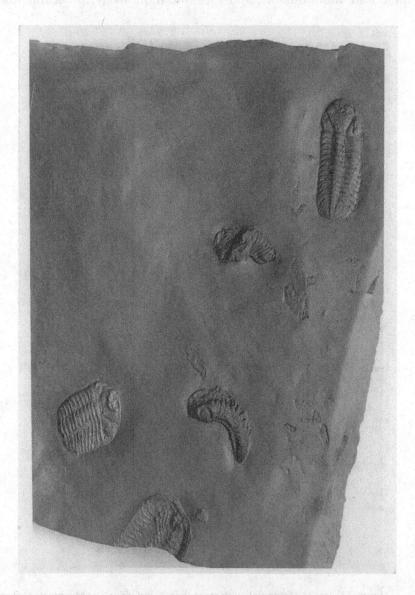

Figure 33 Five pyritized specimens of the trilobite *Chotecops* sp. preserved in different orientations to bedding as a result of turbulent transport, Wingertshell Layer, Eschenbach–Bocksberg mine, Bundenbach (×0.5; HS 414).

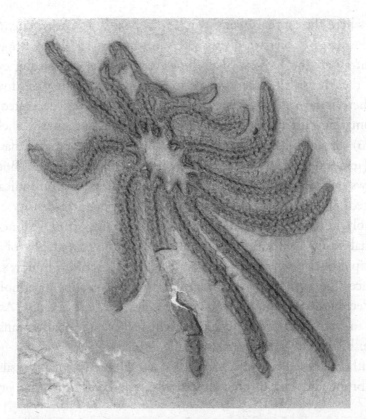

Figure 34 *Medusaster rhenanus* Stürtz, 1890, buried at the top of a graded bed 1 cm thick, Eschenbach–Bocksberg mine, Bundenbach (× 1.2; HS 174).

The occurrence of Konservat-Lagerstätten horizons in the Hunsrück Slate with well-preserved articulated fossils is the result of rapid burial. Complete crinoids in groups that include individuals at different stages of development were clearly killed and buried by the same influx of sediment. Complete arthropods (Figs. 32, 33), asteroids and ophiuroids (Figs. 23, 34) occur in a variety of attitudes to bedding. This too is the result of burial by a sediment-laden density current, but after entrainment and transport. The preserved configurations vary because subsequent collapse and compaction was constrained within the enclosing sediment (as in the case of the fossils of the Cambrian Burgess Shale: Whittington 1971, see Briggs *et al.* 1994).

Previous authors (Koenigswald 1930*a*) interpreted the exceptional preservation of the Hunsrück Slate fossils in terms of transport to a hostile environment favourable for preservation. Models involving transport from a living to a death environment reflect those used to explain the preservation of the Burgess Shale. There preservation is the result of transport by turbidity currents from a pre-slide environment where the animals were living, to a post-slide environment where they became fossilized (Conway Morris and Whittington 1979, see Briggs *et al.* 1994). In contrast to the Burgess Shale, however, there is compelling evidence that many of the Hunsrück Slate animals were living where they are fossilized. In addition there is no evidence that the Hunsrück Slate was deposited at the foot of a submarine cliff or shelf, but rather within basins on the shelf itself (Dittmar 1996). Small folds resulting from syndepositional slumping have been observed at Gemünden (Herrgesell 1978) and in the Eschenbach–Bocksberg quarry. Burrows in the uppermost part of the latter slump horizon indicate that it was colonized after the slumping event. Both slumping and density currents were presumably initiated on the margins of swells, transporting sediment into the depositional basin. Deposition was controlled by the topography of the sea floor so that rapid burial was often confined to areas of just a few hundred square metres. In places the currents were sufficient to uproot groups of crinoids and entrain other echinoderms and arthropods (see Figs. 21, 22, 32, 33). Asteroids which appear to have been rolled up into a tight ball were probably transported some distance. Elsewhere burial was essentially *in situ*.

Other authors argued that preservation of the Hunsrück Slate fossils is the result of an abrupt and catastrophic alteration of the conditions that prevailed

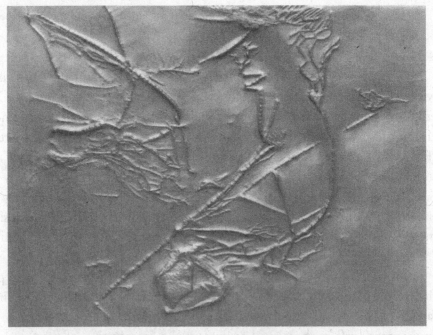

Figure 35 Juvenile crinoids, some of which are attached to a bivalve, Eschenbach–Bocksberg mine, Bundenbach (× 1.0; SNG 083).

where the animals were living. Koenigswald (1930a), for example, invoked the release of hydrogen sulphide or other toxins, including anaerobic water, from sediment or deeper settings. Such scenarios envisage the sea floor as normally inimical to life, and only colonized episodically. Evidence for biological activity, however, including traces produced by a diverse infauna, is ubiquitous in much of the Hunsrück Slate. The occurrence of many juveniles (Figs. 26, 35, 185), some clearly buried *in situ*, indicates that communities of animals were well established. Thus the water column, at least, must have been oxygenated. Oxygen is also a prerequisite for pyritization. It is clear, however, that anaerobic conditions prevailed from time to time within the uppermost layers of the sediment. This is indicated by the formation of pyrite, and the occurrence of horizons with abundant phosphatic concretions.

Lehmann (1957) suggested a different type of basin-wide catastrophe. He argued that volcanic events eliminated life on several occasions during the history of the Hunsrück Slate sea. Although volcanic ash and spilite are known from the sequence (Bartels and Kneidl 1981, Kirnbauer 1986) there is no indication that they had any significant impact on the fauna. Indeed there is no evidence of any widespread mass mortality within the Hunsrück Slate.

It is clear that the burial events that preserve the Hunsrück Slate fossils were local in extent (see Figs. 22, 23). They were a normal aspect of a wider environment where bottom life continually flourished. Their distribution reflects variable relief on the sea floor within the depositional basin.

Pyritization

The importance of the Hunsrück Slate fauna rests largely on the remarkable fossils in the area around Bundenbach and Gemünden where certain horizons (Konservat-Lagerstätten) are characterized by the preservation of soft tissues in pyrite (see Fig. 16). Such preservation is advantageous because it allows the fossils to be examined using x-radiography. Pyrite absorbs x-rays while they penetrate the mudstone matrix. Thus features concealed by the matrix can be revealed. The density of the image reflects the extent of pyritization (see Fig. 22). Although pyrite is ubiquitous in fine-grained muddy sediments, and commonly occurs in association with shelly fossils (Hudson 1982), examples of the pyritization of soft tissues are rare in the fossil record. The only other extensive occurrences of this most distinctive type of soft tissue preservation are in Beecher's Trilobite Bed in the Upper Ordovician of New York State (Briggs *et al.* 1991) and at La Voulte-sur-Rhône in the Middle Jurassic of south-eastern France (Wilby *et al.* 1996).

Figure 36 Pyrite infilling of a limb of the pycnogonid *Palaeoisopus* showing framboids and cubic crystals.

The extent of pyritization of Hunsrück Slate fossils varies. In the Eifel (see Figs. 18, 83, 84), and the Middle Rhine and Taunus regions (see Fig. 119), it is not the main mode of fossil preservation. Here the original calcium carbonate shells of brachiopods, bivalves and trilobites have often been replaced by silica (SiO_2). Where pyrite occurs it is present only as a thin veneer on the surface. Only in the area around Bundenbach and Gemünden in the Middle Hunsrück are soft tissues extensively pyritized. Here too some skeletons are simply covered by a veneer (see Figs. 160, 170), whereas the shell or exoskeleton of others is replaced by a groundmass of pyrite crystals less than 20 µm in dimension together with occasional framboids (Fig. 36) and larger euhedral crystals. Decalcification affects both aragonite and calcite, and is typical of the shells of trilobites, brachiopods, bivalves, ammonites, and orthocone nautiloids, the last often preserved only as a thin pyritized outer periderm. This may reflect the production of acids during decay and pyritization, or undersaturation of calcium carbonate in the sediment. Sometimes only a diffuse imprint of the shell remains (see Fig. 68). This type of preservation is characteristic of specimens of the small bivalve *Buchiola*, for example, from the Kaisergrube in Gemünden.

Although pyrite preserves the form of soft tissues in fossils from the Hunsrück Slate and elsewhere, details of the microstructure do not survive. Such high fidelity preservation is only associated with apatite (calcium phosphate), which replicates cellular details of muscle and other tissues from a range of other

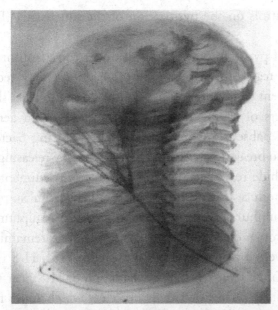

Figure 37 A specimen of *Chotecops* sp. with pyritized appendages and traces of the stomach. The trilobite overlies a small specimen of *Parisangulocrinus* sp., Eschenbach–Bocksberg mine, Bundenbach, the outline showing evidence of tectonic distortion (×0.9; HS 259, radiograph WB8).

Konservat-Lagerstätten (Martill 1990, Briggs *et al.* 1993, Wilby and Briggs 1997). The pyrite that preserves the limbs of the Hunsrück Slate trilobites and other arthropods grew within the cuticle as the muscle decayed (Fig. 37). This process must have happened quickly, before the limbs collapsed and the cuticle itself, which is not present in the fossils, was broken down. The pyrite that preserves some of the soft parts of the fossils from the Jurassic of La Voulte has replaced some of the more rapidly formed apatite which replicated the tissues in the first instance (Wilby *et al.* 1996). The possibility that the soft tissues in the Hunsrück Slate were likewise mineralized in apatite before becoming pyritized remains to be investigated.

In the past, research on the Hunsrück Slate fossils focused mainly on the diversity of animals present rather than the process of preservation. The phenomenon of pyritization received only sporadic attention. It was discussed with reference to the Hunsrück Slate in the early 1930s (Kutscher 1931, Richter 1931) and the mineralogy was investigated some twenty years later (Mosebach 1952*a,b*). Wuttke deduced the processes involved in the pyritization of the Hunsrück Slate fossils based on data from studies of modern sediments (Kott and Wuttke 1987, Bartels and Wuttke 1994). More recently systematic chemical and isotopic analyses of Hunsrück Slate fossils and their enclosing sediments have helped to elucidate the controls on the pyritization of the soft tissues (Briggs *et al.* 1996).

The basic chemistry of pyrite formation is well understood (Allison 1990). It involves a sequence of reactions involving both oxidation and reduction, processes driven by different microorganisms. Pyrite formation is therefore favoured in the upper layers of sediment near the transition from aerobic to anaerobic conditions. In the absence of oxygen, sulphate-reducing bacteria use sulphate in the pore waters to break down organic matter, thereby releasing hydrogen sulphide [1]. This sulphide reacts, in turn, with iron from sediment minerals, mostly iron oxides, to produce iron monosulphide and elemental sulphur [2], or other partially oxidized sulphur species. The partially oxidized sulphur species then oxidize some of the iron monosulphide to pyrite [3]. The remaining iron monosulphide may be converted to pyrite either by reaction with H_2S (Rickard 1996) [4], or with additional sulphur released by oxidation of H_2S near the sediment surface or in the vicinity of burrows. It is not clear which route for converting iron monosulphide to pyrite is the more important.

$$18CH_2O + 9SO_4^{2-} \rightarrow 18HCO_3^- + 9H_2S \qquad [1]$$
$$6FeOOH + 9H_2S \rightarrow 6FeS + 3S^0 + 12H_2O \qquad [2]$$
$$6FeS + 3S^0 \rightarrow 3FeS_2 + 3FeS \qquad [3]$$
$$FeS + H_2S \rightarrow FeS_2 + H_2 \qquad [4]$$

The pyritization of soft tissues, however, only occurs in exceptional circumstances, determined by the distribution and concentration of organic matter in the sediment, and the concentrations of iron and sulphate. Analyses of the matrix surrounding the Hunsrück Slate fossils have shown it to be unusually rich in iron compared with other marine shales (Briggs *et al.* 1996). Furthermore, the concentration of iron around pyritized fossils is higher than that around unpyritized fossils, even within the Hunsrück Slate. This suggests that the concentration of iron may be a significant factor in determining whether or not soft tissues are preserved.

A high concentration of dissolved iron can build up in sediment pore waters in a number of ways, including the oxidation of iron sulphides near the sediment surface, or the bacterial reduction of iron oxides (Canfield and Raiswell 1991). These processes are most prevalent near the sediment surface. Pyrite forms in the zone of sulphate reduction. A high dissolved iron concentration can only be sustained, however, where sulphate reduction is suppressed; otherwise the iron is immediately precipitated into iron sulphides. Little sulphate reduction occurs where the background level of organic carbon in the sediment is low, limiting the available substrate for sulphate-reducing bacteria. When a carcass is introduced into sediment of this kind, it is the focus of a frenzy of sulphate reduction. The sulphide produced cannot diffuse much beyond the carcass because the high surrounding concentrations of dissolved iron can readily diffuse inwards and convert it to iron monosulphide and pyrite on the spot (Briggs *et al.* 1996). Small patches of fine-grained pyrite that occur in the Hunsrück Slate in the absence of any recognizable fossil may reflect the former position of decaying organic matter. This interpretation of the circumstances leading to the pyritization of soft tissues is confirmed by analyses of sulphur isotopes.

Sulphate-reducing microbes use the lighter ^{32}S isotope in preference to the heavier ^{34}S. As long as ^{32}S is resupplied by diffusion through the sediment from the overlying seawater the isotopic signature of the resulting pyrite remains light (i.e. ^{32}S-enriched). When the system becomes closed to diffusion due to further deposition of fine-grained sediment, for example, the microbes reduce more ^{34}S-sulphate and the isotopic signature is correspondingly heavier (^{32}S-depleted). The sulphur isotope compositions of the pyritized Hunsrück Slate fossils are significantly enriched in ^{34}S compared with the pyrite in the surrounding slate. This indicates that pyrite formation continued much longer in the decaying carcass. Some fossils are overgrown by later euhedral pyrite (see Figs. 17, 41, 118, 197, 207). Yet others are surrounded by ill-defined concretions of calcium phosphate, pyrite and other minerals that presumably also reflect decay-induced mineralization.

A combination of two factors is critical to the preservation of soft tissues: rapid burial and sediment chemistry (Briggs *et al.* 1996). Rapid burial is essential to eliminate the scavengers that thrive in oxygenated conditions. The surrounding sediment must contain relatively low concentrations of available organic matter so that dissolved iron and sulphate can diffuse to the carcass rather than precipitating in the sediment. Unusually high concentrations of iron must be present to promote rapid pyrite formation in association with decaying soft tissues. These factors are consistent with the environmental setting evidenced by the sedimentology and paleontology of the Hunsrück Slate. No major catastrophic event is required to explain the remarkable preservation. The water column was oxygenated, and the carcasses were buried rapidly (see Figs. 32, 33), but not very deeply. It is clear that appropriate conditions prevailed for a long period in the Middle Hunsrück region, particularly in the vicinity of Bundenbach and Gemünden. Very precise conditions, however, are required for the preservation of a decaying carcass, and these vary over short distances (Kott and Wuttke 1987).

Different fossils within a small area, and even different parts of a single fossil, may show variable styles of pyritization (see Figs. 22, 236).

Regional variation in the fauna

The Hunsrück Slate biota was dominated by communities of organisms adapted to episodic sedimentation in muddy bottom conditions (see Fig. 26). Evidence of the fauna relies on a patchwork of mines and quarries. While the picture is incomplete it is clear that Konservat-Lagerstätten (horizons with exceptionally preserved pyritized specimens) are confined to small areas, and that the relative abundance of organisms varies from one locality to another (even within a few tens of centimetres of section as is evident in the Eschenbach–Bocksberg quarry near Bundenbach). Different communities of marine organisms are represented in different areas. Asteroids and ophiuroids (starfishes), for example, are characteristic of the Hunsrück Slate around Bundenbach where environmental conditions suited them, but are very rare in other areas (Kutscher 1979b, Mittmeyer 1978, Simpson 1940). Likewise the large spines of acanthodian fishes are common in the north-western part of the Hunsrück Slate outcrop in the Eifel region, but they are extremely rare around Bundenbach. While a proportion of this variation is clearly taphonomic, it appears to be a true reflection of the occurrence of different communities on the sea floor. The classic Hunsrück Slate faunas are from the Middle Hunsrück region, where taxonomic diversity is highest. Elsewhere faunas are relatively depauperate. The Middle Hunsrück faunas provide a basis for comparison with other parts of the Hunsrück Slate. The Middle Hunsrück, Eifel, Middle Rhine and Taunus, and northern Hunsrück and Mosel regions are distinguished by contrasting faunal contents.

The fossils of the Middle Hunsrück region have been the subject of intensive collecting and analysis for decades. Even though the fauna of this region is well documented, new animals continue to be discovered. The majority of Hunsrück Slate species are known only from localities around Bundenbach, Breitenthal and Gemünden (Schmidt 1934, 1941, Lehmann 1957, 1958a,b, Stürmer, Schaarschmidt and Mittmeyer 1980) in the Middle Hunsrück region. Very few Hunsrück Slate taxa occur exclusively outside this area. The proportions of taxa vary within and between localities, presumably reflecting the original distribution of the communities that inhabited different parts of the sea floor at different times. This variability is most striking in the Middle Hunsrück region suggesting that sea floor here may have been more diverse ecologically (Bartels 1994a, Mittmeyer 1980a).

The Middle Hunsrück region is characterized by the abundance, diversity and large size of asteroids and ophiuroids (Lehmann 1957). About 50 species are known from the localities of Breitenthal, Herrstein, Bundenbach and Gemünden, some of them in considerable numbers, and even as mass mortalities (e.g. *Furcaster paleozoicus*: Figs. 23, 136, 196). All stages of development are preserved, indicating that conditions favoured the completion of the entire life cycle in this environment. The bizarre homolazoans, the most common of which is the mitrate (class Stylophora) *Rhenocystis* (see Fig. 138), also characterize this region. Elsewhere the Hunsrück Slate lacks these echinoderms, except for rare single specimens of asteroids and ophiuroids from the Wisper Valley in the Taunus region on the east side of the Rhine (Schöndorf 1909, Becker and Weigelt 1975). Their absence at other localities is almost certainly original. The preservation of other delicate organisms indicates that if asteroids and ophiuroids had been present they too would have been represented in the collections. Even in the Middle Hunsrück region taxa more characteristic of the Rhenish Normal Facies are present where coarser lithologies occur.

Almost nothing has been published on the Hunsrück Slate fossils of the Eifel region (an exception is Kutscher 1941) but intense collecting has yielded a diverse fauna including specimens comparable to those of the Bundenbach area but lacking preserved soft tissues (see Figs. 18, 120). The north-western Hunsrück Slate region, between Laubach and Mayen in the south-eastern Eifel (see Fig. 4), lies close to the transition to the sandier sediments of the Rhenish Normal Facies (Meyer 1988). This is reflected in a higher proportion of silt and in relatively uniform bottom conditions. Many fragments of crinoids occur, but only a few genera are represented. Corals are diverse, many of the rugosans similar to those from the Middle Devonian limestones of the western Eifel (see Figs. 57, 58). The Hunsrück Slate near Mayen is characterized by abundant spiriferid brachiopods (see Figs. 59, 83, 84) which usually occur in sandier sediments and are very rare in more usual Hunsrück Slate lithologies. The trilobite *Parahomalonotus* is relatively common in the Eifel (Brassel and Bergström 1978), but much rarer in the Hunsrück Slate elsewhere. *Chotecops*, on the other hand, is very rare in the Eifel but is the dominant trilobite in the other regions (Figs. 37, 38). Bedding planes crowded with tentaculitoids are unknown in the Eifel although they occur in the Middle Hunsrück region (see Fig. 79) and in the Wisper valley of the Taunus. This may reflect the greater proximity of the Eifel localities to land. Spines of the large predatory acanthodians are common (Fig. 39). These fishes lived in the water column and were less influenced by bottom conditions than many other elements of the fauna. In general the fauna of the Eifel region is

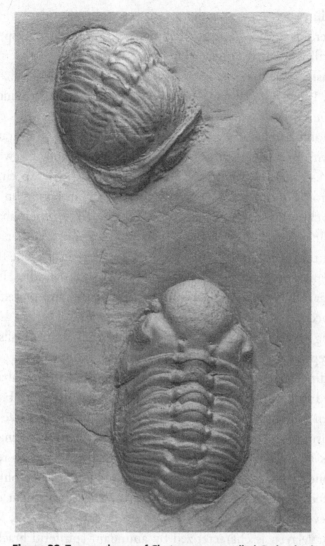

Figure 38 Two specimens of *Chotecops* sp. enrolled, Eschenbach–Bocksberg mine, Bundenbach. This trilobite occurs in nearly all Hunsrück Slate outcrops, but in varying abundance (×0.9, HS 719).

less diverse, much more uniform, and shows greater similarity to the faunas of the sandier Rhenish Normal Facies than to those of the Hunsrück Slate elsewhere.

The Middle Rhine and Taunus regions lie between the Eifel and Middle Hunsrück relative to the margin of the depositional basin to the north-west. At some horizons elements of the Eifel faunas dominate; at others a Middle Hunsrück aspect is more prevalent. Sandstones are commonly interbedded with muddy lithologies, particularly around Geroldstein in the Wisper valley. These

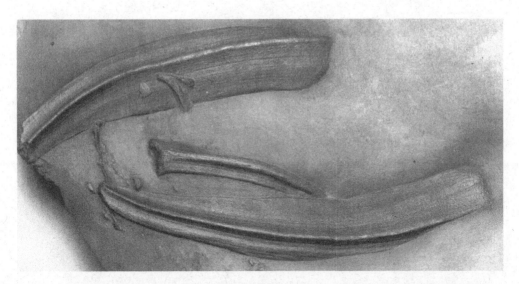

Figure 39 A group of acanthodian spines, Katzenberg mine, Mayen (×0.7; HS 276).

sandstones yield a fauna similar to that of the Rhenish Normal Facies (see Fig. 59). Many of the molluscs listed by Mittmeyer (1980*b*) from the Hunsrück Slate of the Taunus region have been reported only from sandstone beds (Mittmeyer 1978). They do not occur in the normal slate lithologies and are not included in the list of taxa in the appendix.

The faunas of the Hunsrück Slate are poorly documented in the northern part of the Hunsrück mountains and the Mosel valley. Only a few crinoids, rugose corals, orthocone cephalopods and fragments of bivalves are known. More intense collecting would doubtless result in a wealth of new data. Solle (1950), for example, reported a fauna similar to that of Bundenbach from the Gute Hoffnung mine near Altlay (see Fig. 4). The whereabouts of the material is unknown, but the miners confirm that crinoids were present but no starfishes. While body fossils are rare, trace fossils are common at almost all localities in this region. Many were produced by swimming organisms, mainly fishes. Coprolites are also relatively common (see Fig. 220). Thus fishes may have been more abundant in this part of the Hunsrück Slate sea, and the substrate less densely colonized by benthos than elsewhere.

Part II
The fossils

4 Plants

Apart from the marine alga *Receptaculites*, the plants that occur in the Hunsrück Slate are terrestrial, inhabitants of wetlands along the coast or inland, represented by fragments that were transported out to sea (Stürmer and Schaarschmidt 1980). Although only a small number of plant taxa have been discovered, their potential importance far outweighs their numerical abundance because they reveal critical evidence of the nature of the Devonian flora. These fossils have received little attention and would repay detailed investigation.

Far more common than plant macrofossils in the Hunsrück Slate are the tiny spores of primitive vascular plants that can be isolated by dissolving the rock in hydrofluoric acid. These spores could be transported by wind and water over long distances. More than 40 different species have been described (Holtz 1969, Karathanasopoulus 1975). They are potentially useful in biostratigraphy, and their organic maturity indicates that the Hunsrück Slate was subjected to temperatures of about 400 °C in the course of the Variscan orogeny (Wolf 1978, Ecke *et al.* 1985). As the spores are microscopic, and require special techniques to extract them, they are not considered further here.

Algae

Receptaculites

Receptaculitid algae are the only plants that formed part of the benthic communities of the Hunsrück Slate sea (Fig. 40). They secreted a skeleton of interlocking calcareous plates with projections on the inside that strengthened the construction. Only a few specimens of receptaculitid algae are known from the Hunsrück Slate, from the 'Hans' layer in the Eschenbach–Bocksberg quarry near Bundenbach. These examples are unlikely to have been transported and their

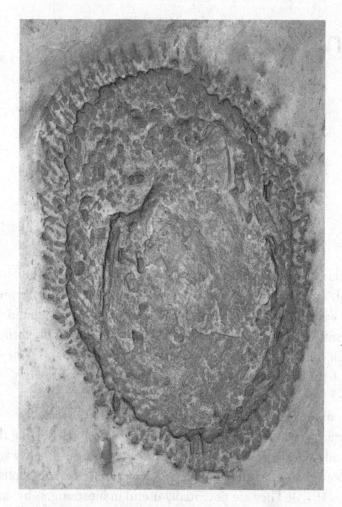

Figure 40 *Receptaculites* cf. *neptuni* (Defrance, 1827), Eschenbach–Bocksberg mine, Bundenbach (×0.7; HS 388).

rarity indicates that receptaculatids only colonized small areas of the Hunsrück Slate sea for limited periods of time. Their requirement for light to photosynthesize constrains the likely depth of the water where they lived to less than 200 m. Although the receptaculitids were clearly algae, the nature of their reproduction is unknown and consequently their affinities cannot be determined (Rietschel 1969, Rietschel and Nitecki 1984). The specimens from Bundenbach have yet to be described, but they are most likely *Receptaculites* cf. *neptuni* Defrance (S. Rietschel personal communication).

Non-vascular plants

Prototaxites

The affinities of *Prototaxites* are enigmatic (Schmid 1976). It is relatively abundant in the Rhenish Normal Facies, and fragments occasionally occur in the Hunsrück Slate. The Hunsrück Slate specimens preserve the branching form, but not the internal morphology (Fig. 41). *Prototaxites* was not a vascular plant. Its habitat is unknown – it may have been coastal or truly terrestrial.

Vascular plants (Tracheophyta)

The Hunsrück Slate has yielded a number of vascular plants (Tracheophyta). Specimens of *Taeniocrada dubia* (?Rhyniopsida), from the Kaisergrube at Gemünden, preserve cellular details (Stürmer and Schaarschmidt 1980). Representatives of the more advanced Trimerophytopsida include *Psilophyton* (Figs. 42, 43), a possible example of *Trimerophyton* (Stürmer and Schaarschmidt 1980), and other undetermined taxa (Fig. 44). The vascular tissue that enabled water to

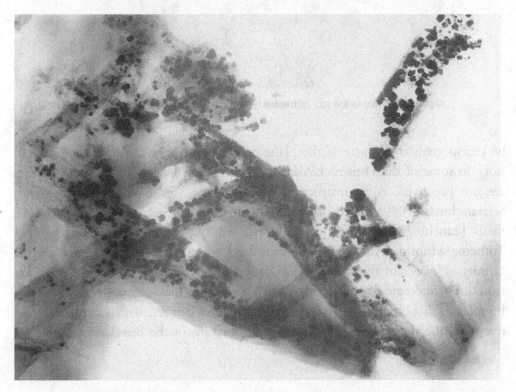

Figure 41 *Prototaxites* sp., showing large late euhedral crystals of pyrite in the interior of the stem, Kreuzberg mine, Weisel, Taunus (× 0.6; HS 309; WS 12330).

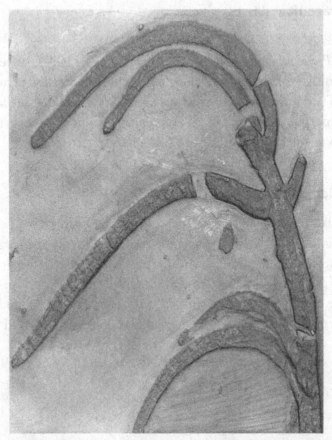

Figure 42 *Psilophyton* sp., Eschenbach–Bocksberg mine, Bundenbach (× 1.0; HS 308).

be transported to all parts of the plant may be evident as a slight longitudinal ridge in some of the Hunsrück Slate specimens, but details of the cell structure are not preserved. A reconstruction of *Psilophyton* (see Fig. 43; Stürmer and Schaarschmidt 1980), based on material from Bundenbach, shows the dichotomous branching, lack of leaves, and terminal sporangia that characterize the Trimerophytopsida. Although these were land plants, they probably grew in the vicinity of lakes and rivers. Fragments of lycopsids (Fig. 45), with small leaf-like structures (microphylls) projecting from a relatively thick stem, are also known. Bundles of dichotomously branched thread-like structures, which may also represent plants, are common at a number of horizons in the Bundenbach area.

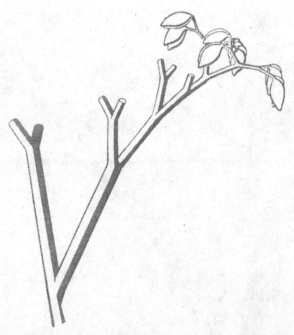

Figure 43 *Psilophyton* sp. Restoration based on specimens from Bundenbach (from Stürmer and Schaarschmidt 1980).

Problematic plants

Maucheria

Rare fragments of a massive stem-like fossil (Fig. 46), known as *Maucheria gemuendensis,* are found at Bundenbach and Gemünden. Ever since this fossil was first described as a plant by Broili (1928*d*) its affinities have been the subject of controversy. The surface of *Maucheria* is commonly reticulate, with small irregularly distributed elevations each surrounding a pore. In places the surface of some specimens is covered by patches of a coarse irregular network which appears to branch and ramify, particularly at its margin. Broili (1928*d*) and Hirmer (1930) dismissed the possibility of an animal affinity for *Maucheria,* whereas Kräusel and Weyland (1930) considered a plant origin equally unlikely! *Maucheria* has received little attention since, and Mittmeyer (1980*b*) listed it as a plant. The specimens may represent stem fragments encrusted by different types of epizoans.

Figure 44 Undetermined large plant (?Psilophytales) of unusually advanced habit for the Lower Devonian. Eschenbach–Bocksberg mine, Bundenbach (× 0.5; HS 527).

Figure 45 Undetermined lycophyte, Eschenbach–Bocksberg mine, Bundenbach (× 1.4; HS 434).

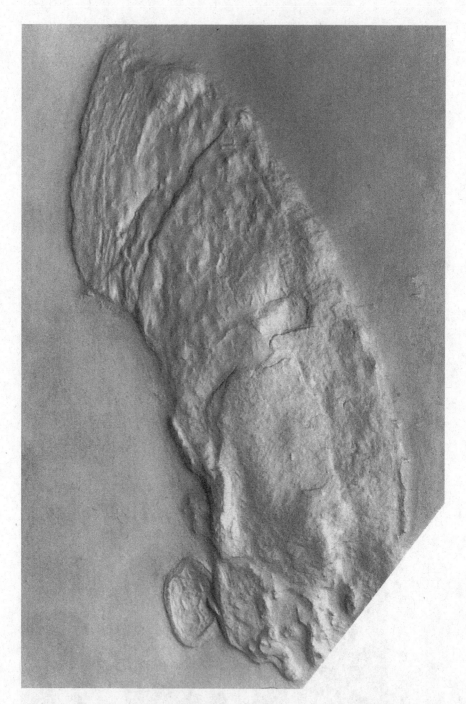

Figure 46 *Maucheria gemuendensis* Broili, 1928, Kaisergrube mine, Gemünden (× 1.1; SNG 229).

5 Sponges to bryozoans

Sponges (phylum Porifera)

The sponges are the simplest type of metazoan represented in the Hunsrück Slate. They consist of aggregates of cells specialized for a relatively small number of functions. They are sessile and feed by filtering. The soft tissues are supported by a skeleton of spicules. Most fossil sponges represent groups in which the spicules are biomineralized and composed of either calcium carbonate or silica. Many sponges, however, have only a proteinaceous skeleton composed of a scleroprotein, spongin.

Sponges are relatively rare in the Hunsrück Slate, but recent observations have revealed significant numbers in some parts of the sequence. Only two species are reasonably well known, *'Protospongia' rhenana* and *Retifungus rudens*. Both belong to the class Hexactinellida, which is characterized by the possession of siliceous spicules. There are, however, a number of other possible sponges which require further investigation. Among these is a single specimen from the Wisper valley (Taunus) which was described as *Asteriscosella nassovica* by Christ in 1925, and several more recently discovered specimens that have yet to be described (including *Retifungus elongatus*, mentioned in Bartels and Brassel 1990, p. 61).

'Protospongia'

This Hunsrück Slate sponge shows similarities to *Protospongia*, which ranges from the Cambrian to the late Devonian. The Hunsrück Slate specimens reveal a bag or tube-like form with a thin outer wall. The spicules were siliceous and the rays projected at right angles. In many specimens they are preserved in pyrite, but in some the spicules are composed of silica (Bartels and Wuttke 1994). These spicules (stauracts) interlocked and supported a rectilinear network of fibres, which are evident on the outer surface of some of the specimens (Fig. 47). Most of the Hunsrück Slate examples of this sponge are fragments (Fig. 48), some up to 1 m in dimension; complete specimens are rare. M. Wuttke and D. Mehl (personal communication) report that this sponge does not belong to the genus *Protospongia*, and the material requires investigation.

Retifungus

Retifungus is known only from the Hunsrück Slate. It was made up of three parts, each characterized by a distinctive arrangement of spicules. The siliceous spicules are covered or replaced by pyrite. The sponge was anchored in the substrate by extremely long parallel diactine spicules, which reached from the root tuft to the oscular opening (Mehl *et al.* 1997). The main part of the body consisted of a

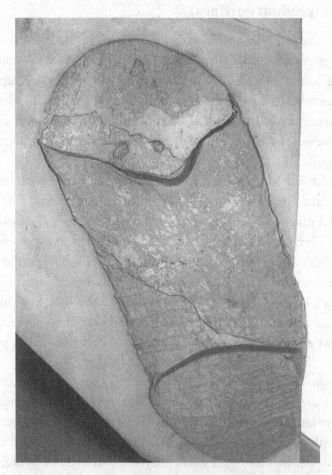

Figure 47 *'Protospongiia' rhenana* Schlüter, 1892. A rare complete individual of this large flask-shaped sponge, Eschenbach–Bocksberg mine, Bundenbach (× 0.5; HS 2).

Figure 48 A fragment of *'Protospongia' rhenana* Schlüter, 1892, showing the spicules, and traces of the network of fibres, Eschenbach–Bocksberg mine, Bundenbach (× 1.0; SNG 218).

long narrow stem of spiralling spicules (Fig. 49). This stem expanded distally into a funnel-like structure comprised of single spicules strengthened by eight bundles of coiled spicules forming a network on the surface. Large diactine spicules were present in the funnel (Fig. 50). The well-preserved specimens allowed this sponge to be restored in detail (Kott and Wuttke 1987; Mehl *et al.* 1997) (Fig. 51). About 15 specimens of *Retifungus* are known of which at least three have crinoids attached to the sponge stem. The group of *Parisangulocrinus minax* illustrated in Fig. 166 may be attached to a *Retifungus* stem.

Cnidarians (phylum Cnidaria)

Chondrophorans (class Hydrozoa, order Chondrophora)

The chondrophorans are a rather enigmatic group of hydrozoans, known as 'by-the-wind sailors'. The most familiar living example is *Velella*. Chondrophorans float on the surface of the oceans buoyed up by gas-filled float chambers that

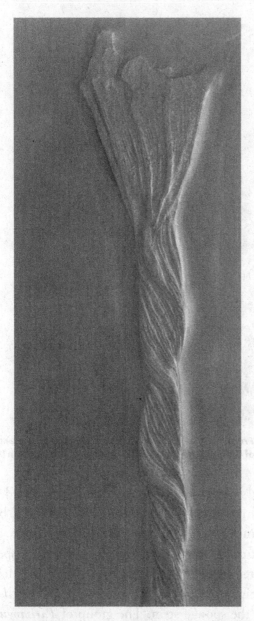

Figure 49 *Retifungus rudens* Rietschel, 1970, showing part of the ropelike stem and the proximal part of the funnel-shaped calyx (× 1.2; SNG 261).

form a disc-like structure called the pneumatophore. The pneumatophore, which is composed of a chitin-like material, bears a vertically projecting median crest which functions as a sail. Below the pneumatophore is a large number of tentacles surrounding the mouth. Some authorities regard the chondrophorans as colonial organisms, others as large, highly modified, individual polyps.

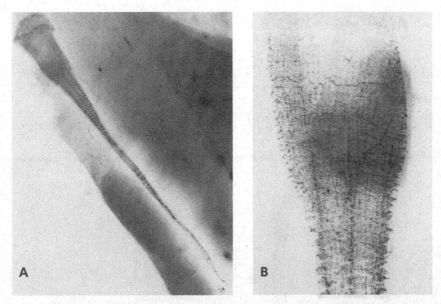

Figure 50 *Retifungus* sp.: (A) a complete animal showing the stem and the funnel-shaped calyx – the x-radiograph shows a specimen of *Ophiurina lymani* Stürtz, 1889, to the right (×0.75, x-radiograph J. Mehl); (B) the proximal part of the calyx showing the arrangement of spicules (×3.0, x-radiograph: J. Mehl).

Plectodiscus

Plectodiscus was originally described by Rauff (1939) but unfortunately his original specimens were lost during the second world war. The new material which formed the basis for a reinvestigation by Yochelson *et al.* (1983) included just one specimen with well-preserved tentacles (Figs. 52–55). This specimen preserves no evidence of the circular disc-like pneumatophore. However, recently discovered specimens (from the Wingertshell Layer, Eschenbach–Bocksberg quarry, Bundenbach) preserve both the tentacles and evidence of the disc (Fig. 53). The tentacles varied in length (and presumably function) and some appear to have been branched. There is evidence that they were curved in transverse section and thickened distally. These Hunsrück Slate specimens are the most completely preserved examples of a fossil chondrophoran yet discovered.

Yochelson *et al.* (1983) also described a number of mainly larger specimens that they interpreted as isolated pneumatophores. These show overlapping concentric structures that correspond to the gas-filled chambers (pneumatocysts) within. A sinuous line evident traversing the disc in some specimens (Fig. 54) was interpreted as the outline of the sail (Fig. 55). Some of these isolated discs are encrusted by brachiopods, bryozoans and other epibionts. Like the pneu-

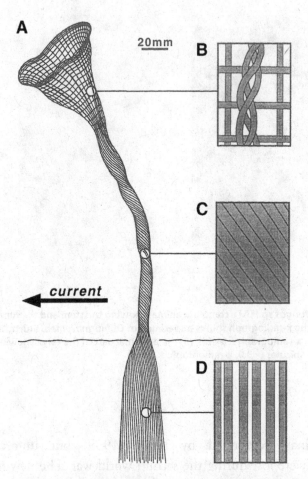

Figure 51 Reconstruction of *Retifungus rudens* Rietschel, 1970: (A) entire sponge showing attitude to water current (×2.5); (B) single and coiled spicules at the base of the calyx; (C) spiralling spicules in the stem; (D) long diactine spicules making up the 'roots'. (From Kott and Wuttke 1987, corrected after Mehl *et al.* 1997.)

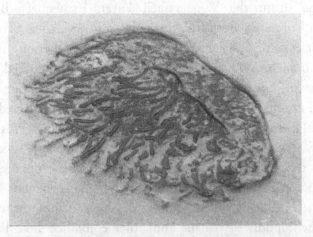

Figure 52 *Plectodiscus discoideus* (Rauff, 1939). Lateral view showing the tentacles, Eschenbach–Bocksberg mine, Bundenbach (×1.0; SNG 265).

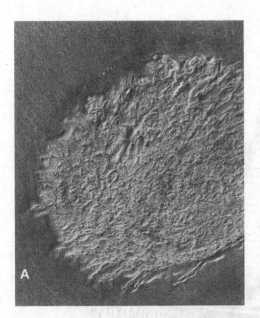
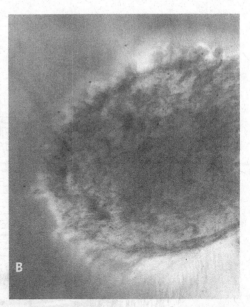

Figure 53 *Plectodiscus discoideus* (Rauff, 1939). Specimen preserving both the tentacles and the pneumatophore: (A) ventral view; (B) x-radiograph showing the outline of the pneumatophore (WB 427). Eschenbach–Bocksberg mine, Bundenbach (× 1.2).

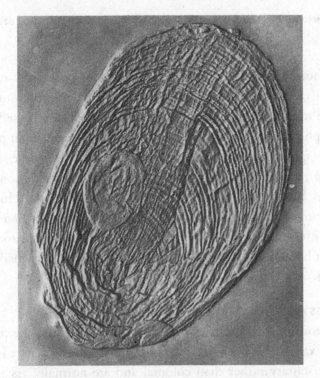

Figure 54 *Plectodiscus discoideus* (Rauff, 1939). Dorsal view of an isolated large disc-like pneumatophore, Bundenbach (× 0.75; SNG 261).

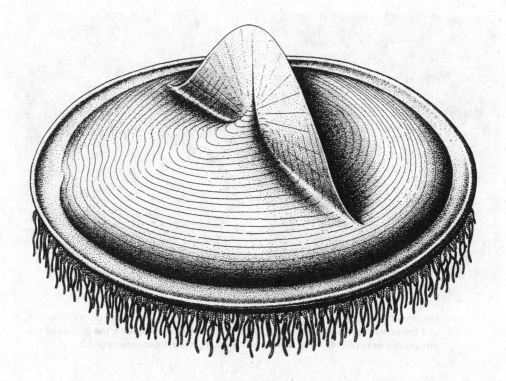

Figure 55 *Plectodiscus discoideus* (Rauff, 1939), restoration (from Yochelson *et al.* 1983).

matophore of living chondrophorans, this must have been the most decay-resistant part of the animal. It clearly survived exposure on the sea floor for some time after the decay of the more labile tissues and before being covered by sediment.

The pneumatophore in the new completely preserved examples of *Plectodiscus* differs in appearance to the larger isolated structures described by Yochelson *et al.* (1983), and their interpretation as the same animal is somewhat equivocal (Bartels and Blind 1995). Otto (1994) argued that all the Hunsrück Slate specimens interpreted as chondrophorines are examples of an *Aspidotheca*-like gastropod! This interpretation had already been rejected by Yochelson *et al.* (1983, p. 66) not least because of the size of the larger discs (which reached diameters of up to 25 cm).

Corals (class Anthozoa)

Corals are very numerous in the Hunsrück Slate, but they have never been investigated in any detail. Examples of both the orders Rugosa and Tabulata are known. The rugose corals are solitary rather than colonial and are normally assigned to the genus '*Zaphrentis*' (Fig. 56). It is clear, however, that a number of rugosans

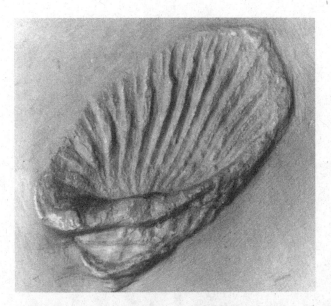

Figure 56 *'Zaphrentis'* sp., Eschenbach–Bocksberg mine, Bundenbach (× 1.6; HS 16).

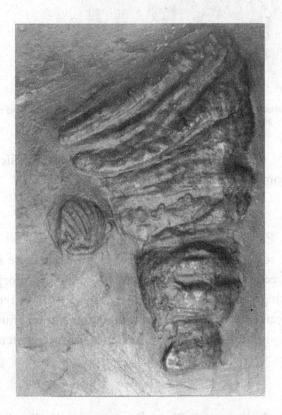

Figure 57 Undetermined rugose coral, Neu Holland mine, Kehrig, Eifel (× 2.0; HS 21).

Figure 58 Undetermined rugose coral, showing rejuvenescence, Neu Holland mine, Kehrig, Eifel (× 1.5; HS 20).

are represented, especially in the Hunsrück Slate of the south-eastern Eifel region (Figs. 57, 58). The most common tabulate coral is *Pleurodictyum*.

'Zaphrentis'

The specimens assigned to *'Zaphrentis'* take the familiar form of a solitary rugose 'horn' coral (Fig. 56). The soft parts are not preserved, but the calcite skeleton is pyritized to varying degrees. The illustrated specimens show the morphology of the coral with its radiating septa. The specimen in Fig. 58 shows rejuvenescence, where growth has been interrupted from time to time and recommenced. This may reflect fluctuations in nutrient supply, or the effect of intermittent burial by the sediment.

Pleurodictyum

The small colonies of the tabulate coral *Pleurodictyum* are typically circular in

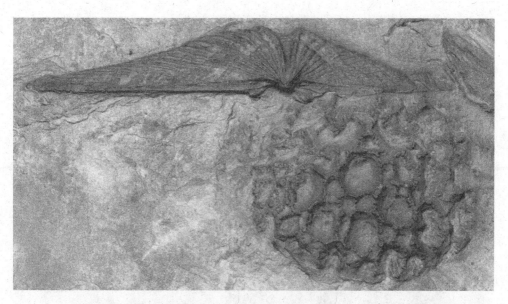

Figure 59 *Pleurodictyum* sp. A small colony clustered with a valve of the large brachiopod *Euryspirifer* sp., and the smaller *Arduspirifer* sp., old roof-slate mine south-west of Geroldstein in the Taunus region (× 1.2; HS 416).

plan. The lower side shows concentric growth lines that give it a wrinkled appearance. The upper side of the colony reveals the polygonal outline of the corallites, which is often hexagonal (Fig. 59). The walls are perforated by mural pores which connect the corallites. Tabulae, which traverse the corallites of tabulate corals horizontally, are usually absent in *Pleurodictyum*. This gives the preserved colony the appearance of a series of cavities.

Pleurodictyum is relatively abundant in the eastern Eifel but rare around Bundenbach. Five species have been recorded from the Hunsrück Slate: *P. giganteum*, *P. hunsrueckianum*, *P. lenticulare*, *P. minimum* and *P. problematicum*. The last of these is abundant in the more sandy sediments of the Rhenish Normal Facies.

Aulopora

The tabulate coral *Aulopora* is very rare in the Hunsrück Slate. This may reflect the lack of suitable substrates, as the colonies usually adhere to a hard surface. The illustrated specimen (Fig. 60) is attached to a fragment of a conulariid. The colony consists of a creeping network of corallites, each expanding to a circular calyx, which is raised slightly above the rest of the skeleton. The network of branching chains of the coral has covered most of the available space on the conulariid. *Aulopora* is abundant in the Middle Devonian limestones of the Eifel region.

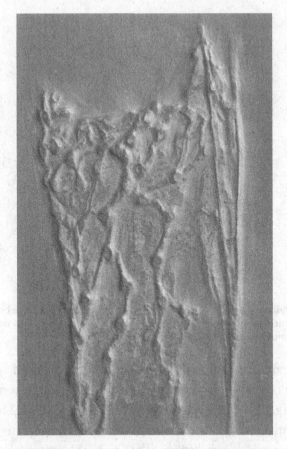

Figure 60 *Aulopora* sp., encrusting a fragment of a conulariid, Eschenbach–Bocksberg mine, Bundenbach (× 1.3) (SNG 233).

Conulariids (class Scyphozoa, order Conulariida)

The extinct conulariids were sessile organisms (except perhaps as juveniles) with a four-sided pyramidal exoskeleton which is all that is normally preserved. They ranged in age from the Cambrian to the Triassic. Conulariids are rare in the Bundenbach area but are more abundant in the slates of the south-eastern Eifel region. Only one specimen has been reported from the Weisel/Taunus region. Although the conulariids are placed here in the class Scyphozoa (following Benton 1993) their affinities have been the subject of debate for over 150 years. Most authors who consider them to belong to a living phylum assign them to the cnidarians, designating them a distinct order Conulariida with closest affinity to the Scyphozoa (see Van Iten 1991). Others regard them as a distinct phylum-level group (see Babcock 1991). Although the Conulariida has become a kind of 'waste-basket' repository for a wide range of forms, some certainly unrelated, they include a clearly defined group characterized by a small number of attributes (Figs. 61, 62): a four-sided phosphatic exoskeleton with a groove at each corner, paired transverse strengthening rods that met in the midline of each of the four sides, and attachment to the substrate by an apical stalk. A case can be made for eliminating forms that lack these characters from the Conulariida (e.g. Babcock 1991). Conulariids are differentiated on the basis of skeleton shape and morphology. Species are distinguished particularly on the nature and spacing of the strengthening rods, which commonly bear nodes and spines, and the style and angle of their articulation. The arrangement of the rods generates a characteristic pattern which is evident through the overlying thin integument.

In some species of Conulariida the four walls extend into triangular or rounded lobes at the aperture (see Fig. 61). Specimens where these terminations are folded in have prompted suggestions that they may have been used to close the aperture, allowing the animal to withdraw completely into the protection of

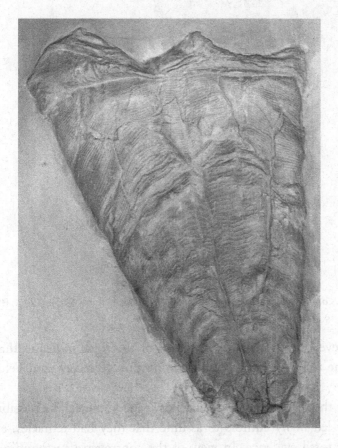

Figure 61 *Conularia hunsrueckiana* Hergarten, 1994, Eschenbach–Bocksberg mine, Bundenbach (×0.8; HS 510).

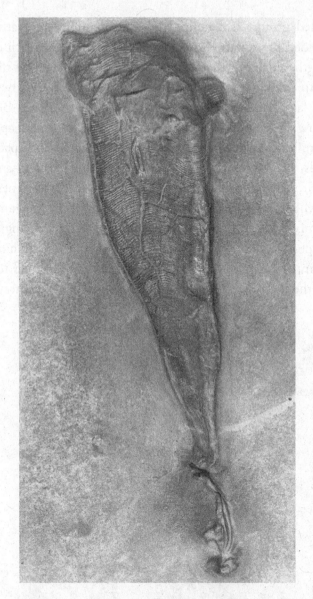

Figure 62 *Conularia bartelsi* Hergarten, 1994, Margaretha mine, Mayen (× 1.1; HS 8).

the skeleton. However variation in the position of the folds indicates that they are taphonomic, the result of collapse and compaction (Babcock and Feldmann 1986).

Knowledge of the soft tissues of conulariids is incomplete. The possibility of a cnidarian affinity led many authors to assume that they had tentacles, even in the absence of any fossil evidence. In view of the controversy surrounding their systematic position, it is hardly surprising that any conulariid preserving an indication of soft parts has attracted considerable attention. Three-dimensionally preserved material from the Lower Carboniferous of the USA revealed evidence of a single elongate tube, presumably a blind alimentary tract, and a number of poorly defined globular structures near the aperture (Babcock 1991). More exciting was the description by Steul (1984) of pyritized soft tissues of conulariids from the Hunsrück Slate, based on x-radiographs. She identified a number of internal organs in several specimens including large eyes with lenses, a stomach, intestine, heart, an 'axial element' and traces of several muscles. She argued that the overall symmetry was bilateral and concluded that the conulariids were chordates.

X-ray examination (by B. Hergarten and J. Mehl) of 11 well-preserved conulariids in the Bartels Collection (German Mining Museum, Bochum) from Bundenbach and the Eifel region revealed no evidence of any of the internal structures identified by Steul. Conversely radiographs of specimens of wide, short orthocone nautiloids from the Hunsrück Slate, similar in outline to conulariids, do show internal structures that recall those described by Steul, in particular an axial element which is clearly the siphuncle. Examination of the specimen (in the Bundenbach Fossilien Museum) upon which Steul based her restoration of most of the internal organs of conulariids revealed that it is clearly an incomplete cephalopod. Thus it seems probable that Steul included a number of orthocone nautiloids among the material that she identified as conulariids (the two can appear very similar when the surface of the skeleton is obscured), leading to a composite and erroneous restoration (Bartels and Wuttke 1994).

Hergarten (1994), who favours a cnidarian affinity for the conulariids, recently investigated the examples from the Hunsrückschiefer, increasing the number of species described from four to nine. *Conularia hunsrueckiana* Hergarten, 1994 (see Fig. 61), *C. gemuendina* R. and E. Richter, 1930, *C. bundenbachia* R. and E. Richter, 1930, *C.* cf *bundenbachia*, *C. tulipina* R. and E. Richter, 1930, *C.* cf *tulipina*, and *C.* cf *mediorhenana* are known from Bundenbach. Four species are known from the Eifel region: *C. mayenensis* Hergarten, 1994 from Mayen, *C. bausbergensis* Hergarten, 1994 from Kehrig, *C. bartelsi* from Trimbs (see Fig. 62), and a new genus *Sinusconularia blasii* Hergarten, 1994 from Mayen. [Hergarten (1994) assigned the new genus and species *Sinusconularia blasii* to the family Circonulariidae Bischoff, 1978 based largely on the supposed circular cross-section. The nature of the cross-section must be in doubt based on a single flattened specimen. The relationship of the Circonulariidae to other conulariids is uncertain (e.g. Babcock 1991).] One species has been described from the Weisel/Taunus region: *C. mediorhenana* Fuchs, 1915. The variable effects of decay and compaction among Hunsrück Slate taxa suggest that further investigation might reduce the number of conulariid species, some of which are based on very few specimens.

Ctenophores (phylum Ctenophora)

The ctenophores were formerly grouped with the cnidarians in the Coelenterata. They are now considered by most authorities to be a separate phylum. Modern ctenophores live in the plankton, and are commonly called comb jellies or sea gooseberries. They do not possess stinging cells, unlike the more familiar jelly-fish, but capture small animals using mucous secreted by cells on both sides of two long tentacles which can be retracted into sheaths. Ctenophores swim by vibrating small flaps arranged in eight rows (the combs) on the body surface, unlike jellyfish which use rhythmic contractions of the bell. They catch prey by swimming slowly trailing the long tentacles and their lateral branches behind in the water.

Apart from the two Hunsrück Slate specimens the only other fossil ctenophores known are three genera based on much better preserved material from the Middle Cambrian Burgess Shale (Conway Morris and Collins 1996). The transparent gelatinous tissue of ctenophores decays readily and, because of their preferred habitat, they are not susceptible to rapid burial. Thus it is not surprising that they are rarely preserved as fossils. Both the Hunsrück Slate ctenophores are based on just a single specimen which was discovered by x-radiography and is too delicate to be revealed by preparation. In view of this, and the uneven nature and variable fidelity of pyritization, their identity remains somewhat equivocal.

Paleoctenophora

This was the first fossil ctenophore to be described (Stanley and Stürmer 1983). The single known specimen was detected by Stürmer when x-raying a slab containing a very large fragment of a cephalopod. The radiograph reveals some of the combs and traces of the long tentacles as well as the statocyst, a balancing organ, which is preserved as a small depression on the top (Fig. 63). The body symmetry and the nature of the tentacles indicate that this fossil represents a ctenophore belonging to the order Cydippida of the class Tentaculata. The restoration (Fig. 64) is based closely on the modern *Pleurobrachia* which is relatively common in the North Sea. Otto (1994) argued that the specimen is a brachiopod, but he did not attempt to reinterpret the diagnostic characters that indicate an affinity with the ctenophores.

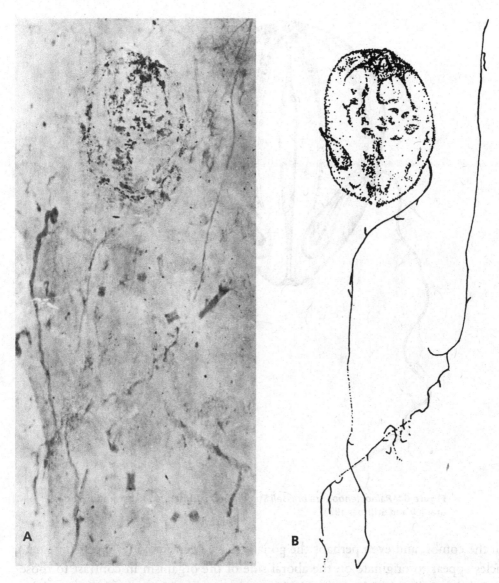

Figure 63 *Paleoctenophora brasseli* Stanley and Stürmer, 1983, Bundenbach: (A) radiograph, WS 1021; (B) interpretative drawing (from Stanley and Stürmer 1983). (× 2.5; Munich, Bavarian State Collection, BSP 1938 ROEM 134.)

Archaeocydippida

This second ctenophore was also discovered by Wilhelm Stürmer during x-ray examination of a slab, and described four years after the first (Stanley and Stürmer 1987). This example shows more detail than *Paleoctenophora*. Structures interpreted as the mouth, the paired sheaths into which the tentacles were withdrawn, and the rows of combs, are evident. Features that may represent the delicate flaps

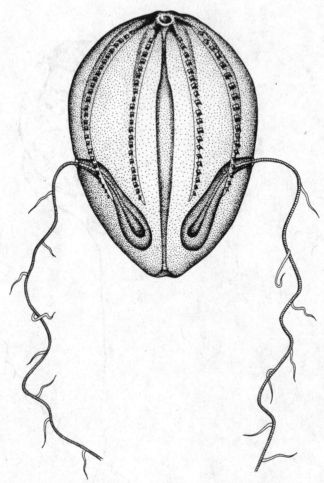

Figure 64 *Paleoctenophora brasseli* Stanley and Stürmer, 1983, restoration (from Stanley and Stürmer 1983).

on the combs, and even perhaps the gonads, have been noted (Fig. 65). The tentacles appear to originate on the aboral side of the organism in contrast to those in *Paleoctenophora* which are attached near the mouth. However, this ctenophore also appears to belong to the order Cydippida. Otto (1994), nevertheless, dismissed this fossil as an artefact of preservation, perhaps not even of organic origin. Although the number of comb rows in *Paleoctenophora* is difficult to discern, there appear to have been eight in *Archaeocydippida* in contrast to the much larger number (up to 80) in the ctenophores from the Cambrian Burgess Shale (Conway Morris and Collins 1996). Thus the number of comb rows may have been reduced to that in modern ctenophores by the Devonian.

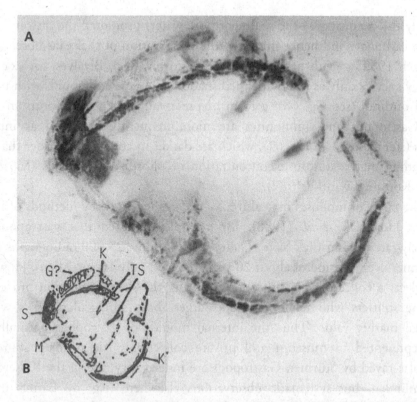

Figure 65 *Archaeocydippida hunsrueckiana* Stanley and Stürmer, 1987, Kaisergrube mine, Gemünden: (A) radiograph, WS 12150; (B) interpretative drawing (from Stanley and Stürmer 1987). The features were tentatively interpreted as follows: K, K', individual comb rows; G?, gonads; TS, two tentacular bulbs; M, mouth; S, sheet-like material adhering to the oral surface, may be degraded muscle fibres or some extraneous material. (× 5.0; Munich, Bavarian State Collection, BSP 1986 I3.)

Molluscs (phylum Mollusca)

The molluscs are one of the most diverse of animal phyla. Fossil representatives are common, particularly in shallow marine settings, from the late Precambrian onwards. Molluscs are not, however, a dominant element of the Hunsrück Slate fauna where echinoderms (particularly crinoids and starfishes) and arthropods are more common. They are nevertheless more important than is generally realized and have been, to a degree, overlooked in favour of the more spectacular pyritized fossils.

Molluscs are normally represented in the Hunsrück Slate, as elsewhere, by their biomineralized shells; the soft parts are rarely pyritized. The relatively low numbers of molluscs in the Hunsrück Slate faunas can be attributed to two factors. Firstly the soft muddy bottom of the Hunsrück Slate sea was unsuitable for many benthic forms, in contrast to the coarser grained Rhenish Normal Facies.

Secondly the chemistry of the sediment pore waters promoted the dissolution of calcium carbonate and hence inhibited the preservation of shells (as observed by Mosebach 1952*b* – with particular reference to bivalves). Bivalves, for example, normally survive only as poorly defined shadow-like external moulds with poorly defined outlines (see Fig. 68), a form of preservation that also occurs in trilobites. Gastropods and ammonites are more frequently preserved as internal moulds (steinkerns) (see Fig. 67), which are darker in colour and harder than the surrounding matrix due to a concentration of phosphatic minerals (Mosebach 1952*b*, Kott and Wuttke 1987).

The rarity of molluscs may also have been exaggerated by methods of investigation. Houbrick *et al.* (1988), for example, reported that gastropods are extremely rare in Hunsrück Slate. Their survey of some 25 000 radiographs made by Stürmer over a period of about 20 years revealed only 10 specimens. However, this reflects a collecting bias introduced during the slate-production process by the slate splitters who retain only the larger more spectacular fossils with a potential market value. Thus the internal moulds of gastropods naturally are underrepresented in museum and private collections, the primary source of material x-rayed by Stürmer. Gastropods are indeed very rare in the Kaisergrube at Gemünden, but systematic observations elsewhere (e.g. by Bartels in the Eschenbach–Bocksberg quarry near Bundenbach) reveal horizons with large numbers of, albeit poorly preserved, gastropods (see Fig. 67). Those horizons occur in the Hans layer of the Bundenbach region, which yields abundant skeletal fragments of corals, orthocone cephalopods, trilobites and crinoids, small fossils that do not attract the attention of either fossil collectors or mineworkers. Gastropods are not rare; in several horizons that crop out near Bundenbach turritiform shells (presumably of a single species) are the characteristic fossils.

Incomplete orthocone nautiloids (normally referred to as *Orthoceras*) are one of the most common fossils at nearly all Hunsrück Slate localities. Calcareous deposits were often laid down during the lifetime of the animal within the distal chambers of the shell as an aid to maintaining its orientation while swimming. These give the orthocones a higher preservation potential than other fossils. The surface of the fossils is often rough, reflecting partial dissolution of the aragonite shell, particularly in specimens from the classic Hunsrück Slate localities. Such dissolution is less prevalent in the roof slates of the Taunus mountains and southeastern Eifel region where orthocones are abundant. Nevertheless the surface of the orthocone shell rarely survives.

Three of the major mollusc groups are known from the Hunsrück Slate: gastropods, bivalves and cephalopods.

Snails (class Gastropoda)

There has been no modern taxonomic study of the Hunsrück Slate gastropods. Relatively small (1.5 – 3 cm), high-spired shells are dominant (Fig. 66) and they are rarely pyritized. The pyritization of internal features was reported by Houbrick *et al.* (1988) who described specimens of four different shell forms. Traces of the kidney, digestive organs, gonads and a possible radula were identified. The outlines of the internal structures are not particularly clear, however, and they could only be interpreted by comparison with the *position* of organs in similar modern forms.

Houbrick *et al.* (1988) argued (on the mistaken assumption that bottom conditions were always anoxic) that the gastropods found in roof-slate lithologies

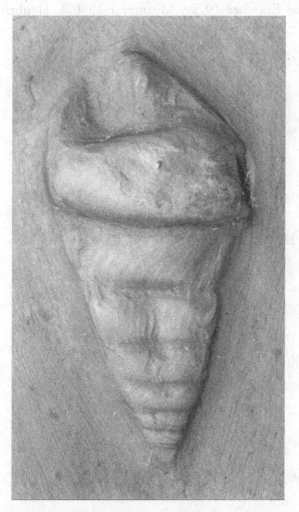

Figure 66 Undetermined gastropod. A pyritized specimen which retains some of the sculpture, Eschenbach–Bocksberg mine, Bundenbach (× 5.0; HS 63).

must have been transported on floating vegetation or lived on tall plants that grew above the anoxic bottom water. The large numbers at some horizons, however, suggest that these gastropods lived on or in the sediment, presumably as deposit feeders. The high-spired morphology may reflect a burrowing mode of life.

Nine genera and 21 species of gastropod have been recorded from the Hunsrück Slate (Kutscher 1979c). Most of these taxa were found in sandstones and quartzites interbedded with the roof slates in the Wisper valley in the Taunus region where they appear to be particularly prevalent (see Mittmeyer 1973, 1980b). Gastropods are less common in roof slate lithologies. In the older literature only *Loxonema obliquiarcuatum* (Fig. 67) was recorded from the Hunsrück Slate at Bundenbach (see Kutscher 1979c, Mittmeyer 1980b). Houbrick *et al.*

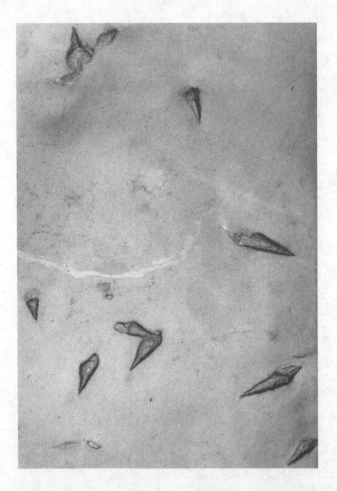

Figure 67 *Loxonema obliquiarcuatum* Sandberger, 1889. Phosphatized specimens showing no preferred orientation, Eschenbach–Bocksberg mine, Bundenbach (× 0.8; HS 59).

(1988) noted that at least seven genera were represented among the 10 pyritized specimens that they discovered but, apart from *Bellerophon*, and forms that resemble *Pycnomphalus* and *Bembexia*, these gastropods could not be determined on the basis of the radiographs. Recently a number of examples of gastropods living on crinoids (in a manner similar to the coprophagous platyceratid gastropods that range from the Ordovician to Permian) have been discovered.

Bivalves (class Bivalvia)

Bivalves are comparatively rare in the Hunsrück Slate. They are normally preserved as distorted external moulds (Figs. 68, 69). Large quantities of tiny *Buchiola* are preserved in this way in the roof slates of Bundenbach and Gemünden. They are pyritized in some horizons in the Kaisergrube at Gemünden. Houbrick *et al.* (1988, p. 401) reported that pyritization is normally confined to a line parallel to the shell margin (perhaps the site of mantle attachment). They discovered examples of only two or three genera in x-radiographs (Fig. 70). Houbrick *et al.* (1988, p. 401) argued that the pyritization of bivalves would have been more extensive had they been buried in life position. They therefore concluded that the specimens were transported. It is more likely, however, at least where they occur in large numbers, that the bivalves lived on the sediment.

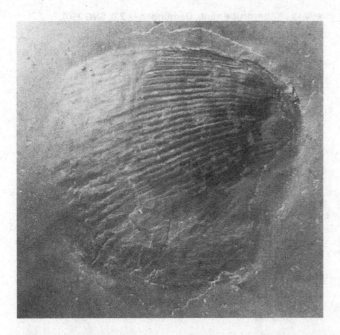

Figure 68 Undetermined bivalve showing partial dissolution of the shell so that the surface detail is lost toward the periphery, Eschenbach–Bocksberg mine, Bundenbach (× 1.3; HS 356).

Figure 69 *Puella* sp., Kaisergrube mine, Gemünden (× 2.0, SNG 250).

Figure 70 *Ctenodonta* cf. *subcontracta* Beushausen, 1895, and *'Zaphrentis'* sp., radiograph, Kaisergrube mine, Gemünden (× 1.6, WS 4588).

A total of 13 species in 8 genera have been recorded from roof-slate lithologies (Kutscher 1966*b*) but they have not been the subject of a modern systematic investigation. As in the case of the gastropods, the list of taxa is much longer if interbedded lithologies in the Taunus region are included (Mittmeyer 1980*b*).

Cephalopods (class Cephalopoda)

Cephalopods are an important element of the Hunsrück Slate faunas. Nevertheless, they have never attracted the interest of either paleontologists or collectors. This, in part, reflects the low monetary value placed on cephalopod fossils, even in the last century, which discouraged their collection by the roof-slate miners and consequently their scientific investigation (Kutscher 1966*b*).

Nautiloids (Subclass Nautiloidea)

Orthocone nautiloids are the most common large Hunsrück Slate fossils. They are particularly abundant in the roof slates of the south-eastern Eifel, for example, where they vary from a few centimetres to more than 0.5 m in length. Most Hunsrück Slate orthocone specimens consist only of parts of the shell (Figs. 71A, 72). It is hardly surprising that the soft tissues which filled the living chamber of these swimming animals normally decayed before they could be replicated by pyrite. Occasionally, however, as in the specimen from Kehring in the south-eastern Eifel illustrated in Fig. 71B, traces of the body and projecting tentacles are preserved extending beyond the chambered shell. Similar structures were reported in cephalopods from the Middle Devonian Wissenbach Slate by Stürmer (1969*b*, 1970*b*) and Zeiss (1969). The siphuncle, the tube which connected the chambers of the shell and permitted the animal to regulate its buoyancy, may also be evident in x-radiographs. This, and traces of other soft-tissue organs of orthocones, were mistakenly attributed to conulariids by Steul (1984; see Hergarten 1994). The orthocone nautiloids may have been preyed upon by large shark-like acanthodian fishes in the Hunsrück Slate sea.

Nearly 20 species of orthocone have been recorded from the Hunsrück Slate (Mittmeyer 1980*b*) but these are all based on taxa described in the last century. The systematics of these cephalopods are in urgent need of attention and most of them are presently only identified as belonging to the genus '*Orthoceras*'. The only investigation of these fossils since the general revision of the Orthoceratidae by Ristedt in 1968 was that of Bandel and Stanley (1989). They identified a number of specimens as lamellorthoceratids. This group of cephalopods is characterized by unique radiating lamellae within the chambers. These organic structures divided the chambers into compartments, controlling the ratio of gas

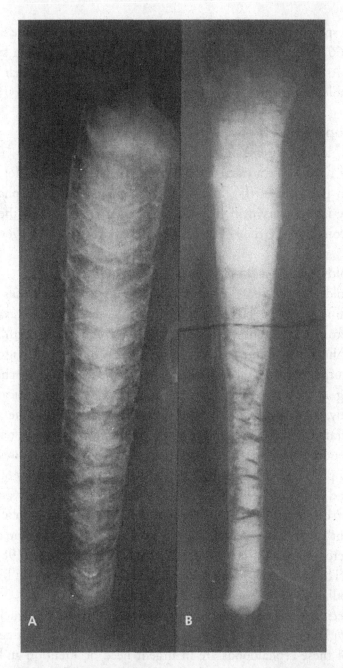

Figure 71 *'Orthoceras'* sp.: (A) proximal end of the chambered shell, infilled with calcite, radiograph, Katzenberg mine, Mayen (× 1.2; HS 47, WS 4969); (B) complete shell showing the soft-parts of the animal within the living chamber – the black lines crossing the shell are cracks infilled with quartz, radiograph, Katzenberg mine, Mayen (× 1.5, WS 4944).

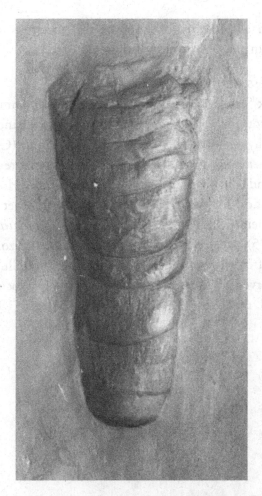

Figure 72 '*Orthoceras*' sp. Proximal end, the shell decalcified and the chambers infilled with sediment, Kreuzberg mine, Weisel, Taunus (× 1.4; SNG 179).

and liquid within the shell. They also provided a template for the precipitation of calcite crusts during the life of the animal. Thus they functioned in controlling the hydrodynamic stability of the cephalopod. The lamellorthoceratids range from Late Siegenian to Eifelian, an interval of about 13 million years.

Bandel and Stanley (1989) assigned the Hunsrück Slate specimens they examined to the genus *Arthrophyllum*. The material came from the Kaisergrube at Gemünden, and from mines in the Mayen area. The Mayen specimens were collected by Bartels and the late G. Blasius, but erroneously ascribed to the Kaisergrube by Bandel and Stanley (1989, Pl. 3, figs. 14–18 [the specimen illustrated in fig. 18 is that in Fig. 31 here]; Pl. 8, figs. 45a–c, 47–49). The diagnostic internal structures of lamellorthoceratids are only evident in x-radiographs or sections. It is not clear whether the shell was external or internal. Bandel and

Stanley (1989) regarded them as an early offshoot of the ancestors of later cephalopods, quite distinct from the orthocone nautiloids.

Ammonoids (subclass Ammonoidea)

The goniatites of the Hunsrück Slate are some of the earliest known ammonites, and they are consequently of considerable significance for our understanding of the evolutionary history and phylogeny of this most important group (Chlupáč 1976). One of the genera, *Anetoceras*, gives its name to early goniatite faunas which occur across the globe. Index fossils are generally very rare in the Hunsrück Slate so these goniatites are critical for dating and correlation with other Lower Devonian strata (Bartels and Kneidl 1981). Some genera, e.g. *Mimagoniatites* (Fig. 73) and *Anetoceras* (Figs. 74, 75), are relatively abundant at some horizons and localities, but the shells have often suffered dissolution (the specimen illustrated in Fig. 73 is relatively well preserved). Besides the ammonites, only Daryoconarida (Tentaculitida) have proved of much use in biostratigraphy (Alberti 1982*a*,*b*, 1983).

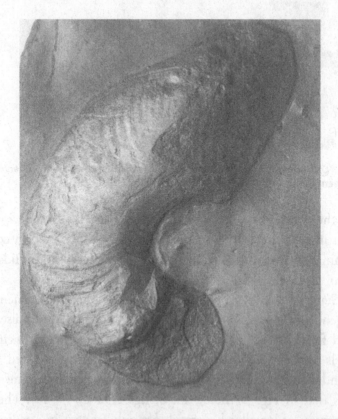

Figure 73 *Mimagoniatites falcistria* (Fuchs, 1907), Kreuzberg mine, Weisel, Taunus (× 1.0; SNG 192).

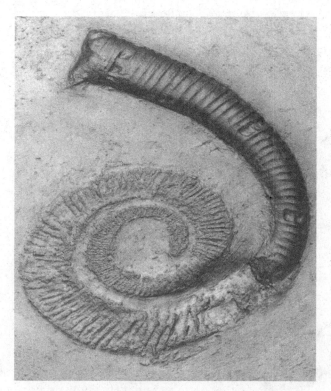

Figure 74 *Anetoceras* aff. *hunsrueckianium* (*sensu* Erben, 1964), Eschenbach–Bocksberg mine, Bundenbach (× 1.0; HS 371).

The systematics of the Hunsrück Slate ammonites were researched most recently by Erben (1953, 1960, 1964, 1965). Twelve species in five genera (*Anetoceras, Gyroceratites, Mimagoniatites, Mimosphinctes, Teicherticeras*) were listed by Kutscher (1969*b*) and Chlupáč (1976). Mittmeyer (1980*b*) listed 13 species in 7 genera (adding ?*Cyrtobactrites* and *Erbenoceras*, which is regarded as a sub-genus of *Anetoceras*).

Coleoids (subclass Coleoidea)

The coleoids are those cephalopods with an internal skeleton; modern examples include the squid, cuttlefish and octopus. The tiny shells of three genera of teuthid coleoids, *Boletzkya, Naefiteuthis* and *Protoaulacoceras* (Fig. 76), occur in the Hunsrück Slate at several localities in the middle Hunsrück area (Bandel *et al.* 1983). The shells were found by careful x-radiography of slabs from the Kaisergrube at Gemünden, emphasizing that intense searching of even well-known strata may yield new taxa. No soft parts were discovered, but the earliest growth stages were preserved in all three taxa. These Devonian forms, like modern coleoids, lacked a larval stage and hatched with an essentially adult morphology.

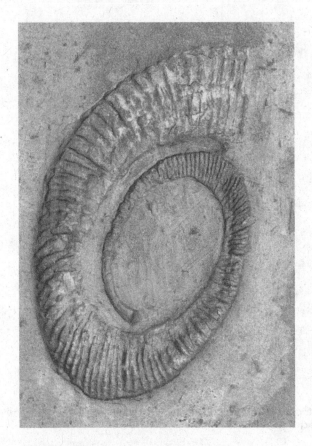

Figure 75 *Anetoceras* cf. *hunsrueckianum* (*sensu* Erben, 1964), Eschenbach–Bocksberg mine, Bundenbach (× 1.6; HS 372).

Stürmer (1985) reported the preservation of soft parts of a new teuthid, based on a radiograph of a specimen from the Schieleberg roof-slate mine near Herrstein in the middle Hunsrück, which he named *Eoteuthis elfriedae* (Fig. 77). In general appearance it shows a strong similarity with the living form *Alloteuthis africanus* from the Gulf of Guinea. The specimen preserves pyritized soft tissues including tentacles, a possible jaw, and a structure which Stürmer (1985) interpreted as a fin. Unfortunately examination of the specimen reveals that the so-called fin is a fracture in the slate (J. Mehl, personal communication). Bandel and Stanley (1989, p. 403) considered it likely that the specimen is not a coleoid but a juvenile nautiloid, probably of the genus *Arthrophyllum*.

Hyoliths (class Hyolitha)

The skeleton of hyoliths consisted of two parts: a long conical shell, the conch, with a flat bottom that accommodated the body, and a small operculum or lid that covered it. A pair of unusual curved appendages, the helens, projected from

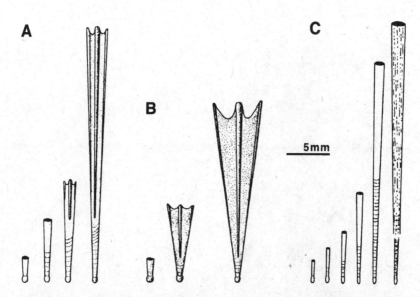

Figure 76 Teuthids from the Hunsrück Slate of the middle Hunsrück region – growth stages of: (A) *Boletzkya*; (B) *Naefiteuthis* and (C) *Protoaulacoceras* (from Bandel *et al.* 1983).

the aperture and may have been used to propel the animal along the surface of the substrate. Hyoliths ranged from the Cambrian to the Permian but Devonian forms are rare. Only two specimens have so far been discovered in the Hunsrück Slate. The systematic position of the hyolithids is controversial (see Runnegar 1980). Some authors consider them to be molluscs, others assign them to a separate phylum.

The illustrated specimen (Fig. 78) shows the conch and the operculum, which has been displaced forward on the right side slightly distorting the symmetry. There is a danger of misinterpreting the soft-tissue morphology based on just one specimen alone. However, a pair of curved structures is evident projecting from beneath the operculum on the left side (the linear structure projecting from a similar point on the right side is a pyritized burrow). These may represent the helens, although Houbrick *et al.* (1988) interpreted them as tentacles. The mass of pyrite within the conch may represent the mantle. The preservation of this hyolith with the operculum and appendages in position indicates that it was buried prior to decay, perhaps in life position. Specimens recently discovered at the Eschenbach–Bocksberg mine at Bundenbach may represent the conchs of hyolithids.

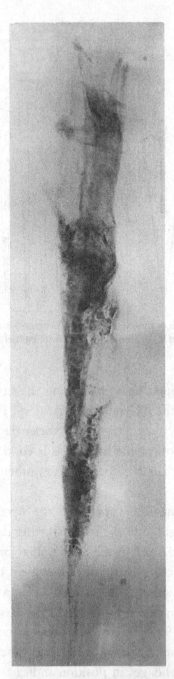

Figure 77 *'Eoteuthis elfriedae'* Stürmer, 1985, specimen interpreted by Stürmer (1985) as a teuthid with preserved soft parts, subsequently identified as a nautiloid, radiograph, Schieleberg mine, Herrstein, Hunsrück (× 2.5; WS 3023).

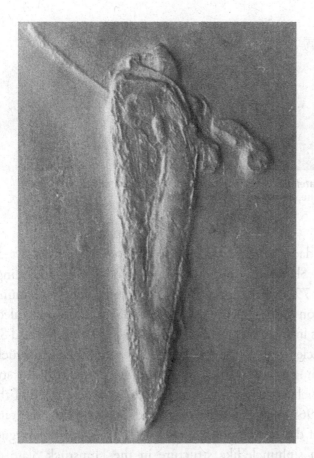

Figure 78 Undetermined hyolith, Eschenbach–Bocksberg, Bundenbach (×2.0; Senckenberg, Brassel collection).

Tentaculitoids (class Tentaculitoidea)

Tentaculitoids are small elongate conical-shelled fossils that ranged from Early Ordovician to Late Devonian. Until the beginning of the 1960s most authors regarded them as a group of pteropod gastropods. Here they are assigned to the molluscs (following Benton 1993) although their affinities have been the subject of some controversy. They are most closely related to either the molluscs (Blind 1969, Blind and Stürmer 1977) or lophophorates (Towe 1978, Larsson 1979). The tentaculitoids that occur in the Hunsrück Slate belong to the order Dacryoconarida which is characterized by a thin shell, an internal surface that reflects the morphology of the exterior, and septa that are poorly developed if at all. Dacryoconarids are useful in biostratigraphy. They have allowed considerable progress in the correlation of the Lower Devonian sediments of the Rhenish and Hercynian zones in Europe (Alberti 1983).

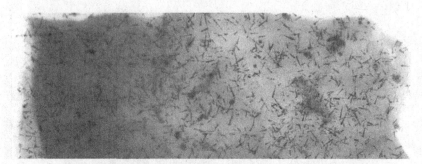

Figure 79 *Viriatellina fuchsi* Kutscher, 1931. A mass mortality, x-radiograph, Kaisergrube mine, Gemünden (× 1.0; WS radiograph).

The soft parts of dacryoconarids are rarely preserved. Brassel collected about 50 slabs of Hunsrück Slate with specimens of dacryoconarids, including some mass mortalities (Fig. 79), on the dumps of the Kaisergrube in Gemünden in 1970. X-ray examination of these slabs (at a 60-fold magnification) revealed traces of the gut and tentacles in some specimens (Brassel *et al.* 1971). Blind and Stürmer (1977) identified tentacles in the aperture of a specimen from the Hunsrück Slate, suggesting a filter-feeding mode of life. They also interpreted a linear structure as evidence of a siphuncle and this, together with the microstructure of the shell revealed by Blind's (1969) earlier study, supported his view that the tentaculitoids were a subclass of the cephalopods. Yochelson (1989), however, argued that the identification of a siphuncle-like structure in the Hunsrück Slate dacryoconarids was based on a misinterpretation of the x-radiographs and considered that the affinities of the group lie with some of the worm phyla.

Tentaculitoids were previously considered nektonic. The shell was interpreted as a buoyancy aid, like that in cephalopods. However structural similarities to other tube-like forms called microconchids, which are cemented to the substrate, raise the possibility that tentaculatids were sessile – the shell standing vertically in soft sediment with the aperture and feeding apparatus extending above it into the water column.

A taxonomic revision of tentaculitoids (Boucek 1964) showed that the two species previously recorded from the Hunsrück Slate are identical. This form is known as *Viriatellina fuchsi* (see Boucek 1964, Kutscher 1979*a*). Many radiographs of Hunsrück Slate slabs, however, revealed large numbers of tentaculitoids, and the variation in shape suggested a greater diversity than one species. This was confirmed by Alberti (1979, 1982*a,b*, 1983) who reported five additional species (including two new ones) of the dacryoconarid *Nowakia* from the Hunsrück Slate of the Bundenbach region. Representatives of a second order of

Figure 80 *Tentaculites grandis* Roemer, 1844 (larger specimens, mainly in upper half of photo) and *T. schlotheimi* Koken, with *Pleurodictyum* sp.and fragments of crinoids; old roof-slate mine south west of Geroldstein, Taunus region (×0.6; HS 89).

tentaculitoids, the Tentaculitida, have been reported from the Taunus region (Mittmeyer 1980*b*) (Fig. 80).

Brachiopods (phylum Brachiopoda)

Brachiopods, although a relatively minor phylum in the seas of today, are some of the most common fossils in Devonian marine deposits. There are two major groups (classes): the Articulata, which have thick calcareous shells, and the Inarticulata, which have thinner calcium phosphate shells. The Articulata are particularly abundant in limestones and shales deposited in shallower water. Thus, as in the case of the molluscs, a much greater diversity of brachiopods is known from the Rhenish Normal Facies in the Taunus region than from the roof-slate lithologies (Mittmeyer 1980*b*).

In the roof slates brachiopods, like bivalved molluscs, are usually at least partly decalcified and preserved as external moulds (Fig. 81). Such dissolution

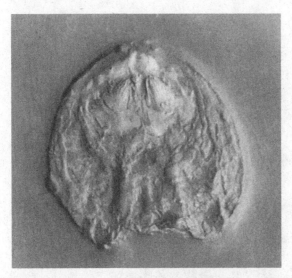

Figure 81 Undetermined brachiopod, Eschenbach–Bocksberg mine, Bundenbach
(× 1.8; SNG 234).

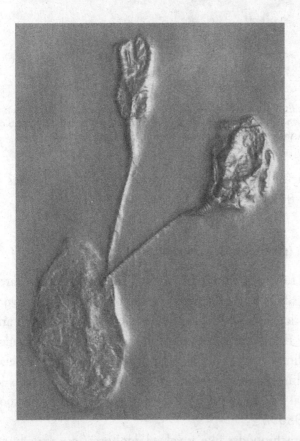

Figure 82 Undetermined shelly fossil, perhaps a brachiopod, with two small *Codiacrinus
schultzei* attached, Eschenbach–Bocksberg mine, Bundenbach (× 1.2; SNG 098).

Figure 83 *Hysteriolithes (Acrospirifer) arduennensis* Solle. A silicified specimen preserving the spiralia, Katzenberg mine, Mayen, Eifel (× 1.3; SNG 241).

Figure 84 *Arcrospirifer arduennensis prolatestriatus* Mittmeyer, a silicified specimen associated with a tabulate coral and other shelly debris, Margaretha mine, Mayen, Eifel (× 0.8; HS 71).

exaggerates their rarity, and makes those that are preserved difficult to identify. It is impossible to determine, for example, whether the shell which serves as an attachment site for the two crinoids (*Codiacrinus*) illustrated in Fig. 82 is a bivalve or a brachiopod. Well-preserved spiriferid brachiopods are common in the Hunsrück Slate of the south-eastern Eifel, however, reflecting a higher sand content in the sediment (Figs. 83, 84). Here the calcite of the shells has been replaced by silica. Occasionally the delicate spiral that supports the lophophore (the structure that gives spiriferids their name) is even preserved (Fig. 83). These brachiopod faunas have yet to be investigated in detail.

Brachiopods are sessile organisms attached to the substrate by a stalk or pedicle. They feed using a lophophore with two arms that project into the mantle cavity, which forms a filter chamber that occupies most of the internal volume of the shell. In the spiriferids the shell has a long hinge which sits directly on the sediment, supporting the shell. Pyritized traces of the pedicle of brachiopods have recently been reported from the Hunsrück Slate (Südkamp 1997).

Bryozoans (phylum Bryozoa)

Bryozoans are small colonial animals that range from the Ordovician to the present day. The individual zooids of the bryozoan colony filter feed using a lophophore, an attribute that demonstrates their close relationship to brachiopods and phoronid worms. The colonies may be free standing on the surface of the sediment or they may encrust surfaces, including other organisms or their remains (Fig. 85). Bryozoans are relatively common in the Hunsrück Slate but, perhaps because of their small size and relatively inconspicuous nature, have received little attention. Only one, *Hederella* sp., has been described in detail (Brassel 1977).

?Fenestella

Specimens tentatively identified as *Fenestella* have been found in the Hunsrück Slate of the Eifel region where the original skeleton may be replaced by silica, but they have yet to be investigated in detail. The branches that support the tiny

Figure 85 Undetermined bryozoan encrusting the surface of an orthocone, Katzenberg mine, Mayen, Eifel (× 1.3; HS 474).

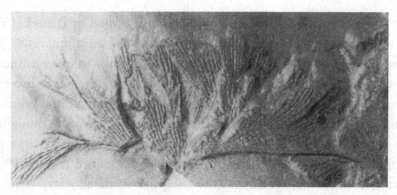

Figure 86 *Fenestella* sp., Bausberg II mine, Kering, Eifel (× 1.2; HS 3).

zooids of this genus are connected by transverse struts or cross-beams known as dissepiments. The most striking feature of the skeleton, therefore, is a large number of apertures or windows (fenestrules) which give *Fenestella* its name. The colonies lived on the sea bed anchored in the sediment. Their normal cup- or fan-shape is obscured by flattening in the Hunsrück Slate examples (Fig. 86).

Hederella

Hederella (Fig. 87) was first recorded from the Hunsrück Slate when a specimen was described by Brassel (1977). The genus is relatively rare in Europe. Most of the species have been described from the Middle Devonian of North America

Figure 87 *Hederella* sp. encrusting a fragment of the shell of *Anetoceras*, Bundenbach (× 1.8; SNG 232).

(Bassler 1939, Solle 1952, 1968). The specimen described by Brassel (1977) preserves two colonies which settled inside the shell of an ammonoid, probably *Anetoceras*. The initial zooids that produced the bush-like colonies cannot be distinguished. The shorter axes in the centre of the colony each gave rise to two or three remarkably long, lateral branches. The axes expanded distally into a club-like termination. The Hunsrück Slate specimens of *Hederella* are thought to represent a new species. The nature of the pyritized preservation does not permit a clear diagnosis, however, and the species has not been named (Brassel 1977). The group to which *Hederella* belongs (the suborder Hederelloidea) is in need of taxonomic revision. Some authors question whether they are bryozoans or more closely related to tabulate corals (see Taylor, p. 486, in Benton 1993).

6   Annelids and arthropods

Annelids (phylum Annelida)

Bristle worms (class Polychaeta)

The great majority of the 25 000 radiographs of Hunsrück Slate slabs in the Stürmer archive at the Senckenberg Museum, Frankfurt, reveal long string or ribbon-like structures preserved in pyrite. The crystals of pyrite that formed in these burrows are often evenly spaced giving the appearance of a series of short elements which may be misinterpreted as fragments of crinoid stems. These burrows occur in several distinct patterns, some of which appear to be characteristic of particular horizons, but they have yet to be studied. The traces presumably represent the activity of a range of different deposit-feeding worm-like forms. The pyrite grew in association with the material that passed through the gut. The degree of pyritization varied depending on the nature of the organic material. An investigation of these traces would cast light on the activity of the fauna in the upper layers of the sediment on the floor of the Hunsück Slate sea.

Although traces of the deposit-feeding activities of worms are almost ubiquitous in the Hunsrück Slate, evidence of the animals that made them is much rarer. Recently, however, a number of specimens of soft-bodied polychaetes have been discovered. They add significantly to our knowledge of Paleozoic polychaetes which, with the exception of the Cambrian Burgess Shale, and the Carboniferous Bear Gulch (Montana) and Mazon Creek (Illinois) deposits, are rarely fossilized (see Briggs and Kear 1993). Genuine worms are difficult to find in the Hunsrück Slate. Fragments of other animals (e.g. sections of crinoid stems, starfish arms) may, following decay and degradation, have a worm-like appearance. Preparation or x-radiography may be necessary to confirm whether or not a fossil is a worm.

The Hunsrück Slate polychaete specimens preserve the outline of the worm, and some three-dimensional relief remains. The body itself appears largely featureless, although a trace of the gut may survive. The most striking structure is the chitinous chaetae or bristles which project from the fleshy limbs (parapodia) that were used in locomotion. The chaetae were presumably more readily pyritized than the rest of the animal due to their resistance to decay (Briggs and Kear 1993). The outline of the soft tissues of the limbs, including their delicate projections or cirri, may also be evident. One of the known specimens preserves a concentration of pyrite in the head but it is not clear what this

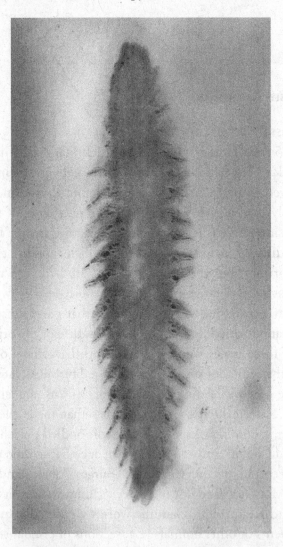

Figure 88 *Bundenbachochaeta eschenbachensis* Bartels and Blind, 1995, radiograph (×2.5, WB 296).

represents. There is no evidence of a jaw. Polychaete jaws are made up of decay-resistant elements which often occur as isolated microfossils (scolecodonts). The absence of a jaw suggests that these polychaetes were deposit feeders and may have made the burrows while feeding.

Bundenbachochaeta

This polychaete, discovered in 1994 in the Eschenbach–Bocksberg mine near Bundenbach, clearly shows the pyritization of soft tissues (Bartels and Blind 1995). The worm is 48 mm long and has 22–24 segments (Figs. 88, 89). Each biramous parapodium shows two unequal bundles of chaetae, one (?the notopodium) comprised of relatively long strong bristles, the other (?the neuropodium) consisting of a fan of tiny short ones. The x-radiograph (Fig. 88) shows the spine-like aciculae extending from the parapodia into the body, serving as an attachment for the muscles used in locomotion. The pyritized aciculae are broken into several pieces. Two specimens are known, the holotype (Fig. 88), and a fragmentary specimen in a private collection in Bundenbach.

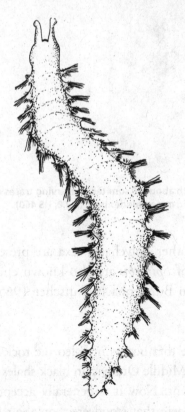

Figure 89 *Bundenbachochaeta eschenbachensis* **Bartels and Blind, 1995, restoration (from Bartels and Blind 1995).**

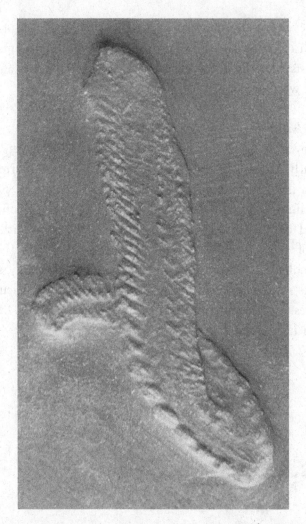

Figure 90 Undetermined polychaete with about 60 somites, preserving traces of the parapodia in relief, Eschenbach–Bocksberg mine, Bundenbach (×2.0; HS 460).

Pyritized soft tissues of at least two other polychaete taxa are preserved in the Hunsrück Slate (Figs. 90, 91). Tubes of *Spirorbis* are also known encrusting examples of *Conularia* and *Maucheria* from Bundenbach (Kutscher 1965b).

Sphenothallus

Sphenothallus is an elongate phosphatic tube that occurs in Paleozoic rocks. It was originally described in the last century from Middle Ordovician black shales in New York State, but was misinterpreted as a plant. Now it is generally accepted that *Sphenothallus* is the tube of a sessile polychaete that stood erect on the substrate. The specimens from the Hunsrück Slate (Fig. 92) are important in that they provide

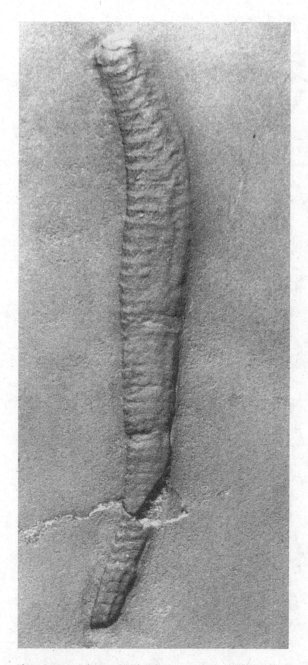

Figure 91 Undetermined polychaete with about 50 somites; parapodia are not preserved but short spines originate at the segment margins, Eschenbach–Bocksberg mine, Bundenbach (× 2.0; HS 284).

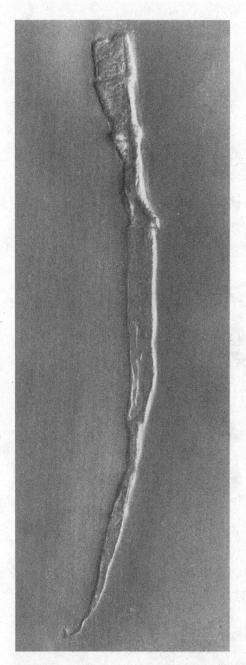

Figure 92 *Sphenothallus* sp. illustrating the form of the tube, Bundenbach (× 1.9; SNG 316).

the only known evidence of the soft parts (Fauchald *et al.* 1986). X-radiographs occasionally reveal a bilaterally symmetrical apparatus of tentacles in the aperture, although details are not clear (Fig. 93). A crown of tentacles may have been extended from the tube for filter-feeding. Fauchald *et al.* (1986) emphasized that the small

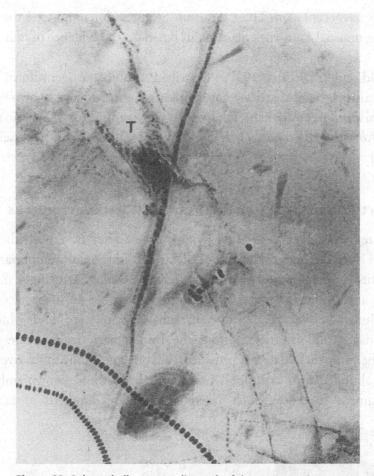

Figure 93 *Sphenothallus* sp. X-radiograph of the aperture of a specimen showing a bilaterally symmetrical structure (T) that may represent the tentacles (top left). The specimen is traversed by a burrow and associated with two crinoid stems. (× 5.0; Munich, Bavarian State Collection, BSP 1986 I 5, WS 515 a).

number of morphological characters in *Sphenothallus* prevent its confident assignment to any annelid family. Indeed even an annelid affinity is equivocal in view of the phosphatic composition of the tube (in contrast to the calcium carbonate found in modern annelids) and the lack of evidence for segmentation of the soft parts.

Misidentified worm-like animals

Fauchald *et al.* (1988) described two worm-like specimens from the Hunsrück Slate, which they compared to a tunicate chordate and a platyhelminth, solely on the basis of x-radiographs. Fauchald and Yochelson (1990*b*), however, noted that the so-called tunicate is a partially pyritized cephalon of the trilobite *Chotecops*. Although they regarded the other specimen as a plausible 'flatworm', it too is a *Chotecops* cephalon. These errors of interpretation emphasize the danger of identifying unusual Hunsrück Slate fossils based on the evidence of radiographs alone.

Fauchald and Yochelson (1990*a*) described a tubicolous vermiform animal from the Hunsrück Slate that they considered to be unlike any known fossil or recent organism. The specimen, however, is an example of the edrioasteroid *Pyrgocystis (Rhenopyrgus) coronaeformis* (see Fig. 174) (Bartels and Brassel 1990, pp. 88, 177).

Arthropods (phylum Arthropoda) – general remarks

Arthropods represent over 75% of all living species. They have been the dominant phylum since the major radiation of metazoans in the early Cambrian. They have diversified to occupy nearly every conceivable ecological niche – in water, on the land, and in the air. The arthropods are among the most spectacular fossils of the Hunsrück Slate. The most remarkable examples are those which preserve appendages and soft tissues from the Bundenbach region. The muddy bottom of the Hunsrück Slate sea provided suitable conditions for a diversity of forms. The abundance of arthropod taxa varies from those known only from unique specimens to those which are represented by hundreds of examples.

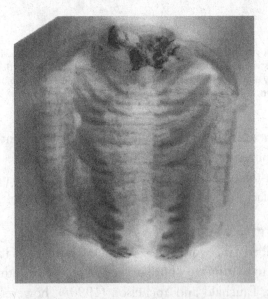

Figure 94 X-radiograph of *Chotecops* sp., the most common arthropod in the Hunsrück Slate, showing the appendages, and the partially pyritized stomach contents in the head region, Eschenbach–Bocksberg mine, Bundenbach (× 1.0, WB 9).

Trilobites are particularly abundant, and of these the phacopid *Chotecops* is the most common (Fig. 94).

Arthropods are characterized by a bilateral symmetry. The body is divided into somites which are grouped into tagmata (e.g. the cephalon, thorax and pygidium of trilobites) with particular functions. The substantial fossil record of the arthropods may be attributed to the possession of an exoskeleton composed of chitin which, in some cases, was strengthened with calcium carbonate (e.g. the dorsal exoskeleton of trilobites). The construction of the exoskeleton may reflect the tagmosis beneath. Thus the pygidium of trilobites, for example, covers several somites. The subdivision of the limbs into articulating segments, which gives the arthropods their name, is a necessary consequence of the stiff exoskeleton. Arthropods increase in size by moulting. A large proportion of fossil arthropods are the remains of moults. The abundance of trilobites therefore probably reflects the high preservation potential of their moulted exoskeleton.

The remarkable x-ray images produced by Wilhelm Stürmer, together with the discovery of new material, provided the basis for redescriptions of most of the arthropod groups from the Hunsrück Slate from 1973 onward. Stürmer's

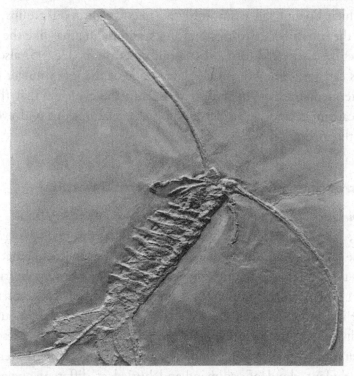

Figure 95 A new undescribed arthropod of unknown affinity, Eschenbach–Bocksberg mine, Bundenbach (× 0.7, HS 574).

work, particularly in collaboration with Professor Jan Bergström of Stockholm, clarified our knowledge of the appendages and soft tissues of many of the arthropods. This research remains one of the most significant contributions to the study of the Hunsrück Slate fauna.

The morphological data preserved in some of the Hunsrück Slate arthropods is comparable to that in specimens from the Cambrian Burgess Shale type faunas (Briggs *et al.* 1994), and the phosphatized Orsten faunas (e.g. Müller and Walossek 1988, Walossek 1993). In contrast with the Cambrian, however, there are very few exceptionally preserved biotas of Ordovician, Silurian and Devonian age. Thus the Hunsrück Slate fauna is a unique source of data on later Paleozoic arthropods. The fauna includes examples of the major aquatic groups, crustaceans, chelicerates and trilobites. Only the terrestrial myriapods, with uniramous limbs, are absent. The preserved soft-part morphology of the crustacean *Nahecaris*, the chelicerate *Weinbergina* and the trilobite *Chotecops* is of fundamental importance to evolutionary studies. The Hunsrück Slate yields the only known fossil examples of adult pycnogonids (sea spiders) so far described (although examples from the Jurassic of La Voulte-sur-Rhône, France, await description). Most importantly, perhaps, the fauna includes a number of forms like *Mimetaster*, *Vachonisia* and *Cheloniellon* which do not fall easily into the major arthropod groups and are reminiscent of the intermediate forms found in Cambrian faunas like the Burgess Shale. Two recently discovered forms that await description (Fig. 95) also appear to fall into this category. The Hunsrück Slate is one of the few younger deposits where such intermediates have been discovered and the evidence that they provide is of critical importance to understanding the relationships and evolutionary history of the arthropods.

Marrellomorphs (Subphylum Marrellomorpha)

Mimetaster

Gürich (1931, 1932) named this unusual looking arthropod *Mimetaster* in response to the star-like outline of the dorsal shield; the species name, *hexagonalis*, refers to the six large radiating spines. It was redescribed by Stürmer and Bergström in 1976. A recently discovered mass mortality of *Mimetaster* within the Hans layer in the Eschenbach–Bocksberg quarry yielded a slab about 25 cm by 30 cm with more than 20 individuals (see Fig. 32). These are preserved in various orientations to the bedding, indicating that the specimens were transported in a turbulent cloud of sediment and buried at different angles in the mud.

The central part of the star-shaped dorsal shield of *Mimetaster* (Figs. 96, 97) covered the head. A pair of eyes projected from the shield on long stalks. There was also a median eye, consisting of a pair of ocelli. Each of the six large spines radiating from the dorsal shield had two rows of slender spines projecting from it laterally, the whole providing a kind of network that covered the animal. The head bore a pair of long, delicate antennae and two pairs of robust walking legs, the first much larger than the second and each consisting of seven segments of varying lengths. There was a large labrum which presumably concealed the opening of the posteriorly facing mouth. In the centre of the labrum, on the outer surface, was a small oval structure (which varies in its preserved morphology) which may have been a sensory organ. The trunk extended between the two large posterior spines of the dorsal shield. It was disproportionately short, but included up to 30 segments diminishing in size to the rear, each with a pair of biramous limbs (Fig. 98).

There have been some unlikely suggestions for the mode of feeding in

Figure 96 *Mimetaster hexagonalis* (Gürich, 1931), ventral view, Bundenbach (× 1.6; private collection).

Figure 97 *Mimetaster hexagonalis* (Gürich, 1931). Two specimens in ventral view showing well-preserved appendages including the filamentous outer branches, Eschenbach–Bocksberg mine, Bundenbach (× 0.9; HS 318).

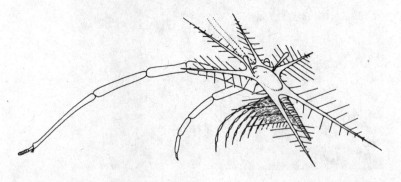

Figure 98 *Mimetaster hexagonalis* (Gürich, 1931), restoration (from Stürmer and Bergström 1976).

Mimetaster: that it used a suctorial mouth to feed on the ophiuroids that are often found associated with it (Gürich 1931, Lehmann 1956*a*), or that the network of spines on the carapace functioned in food capture by filtering (Birenheide 1971). It is more likely, however, that *Mimetaster* was a deposit feeder, supporting itself on the substrate with the large head appendages and posterior spines while using the trunk limbs to stir up the sediment in pursuit of organic particles and small animals. A number of specimens of *Mimetaster* show a mass of sponge material entangled in the dorsal shield. This spiny structure may therefore have been an adaptation to trap material that acted as protective camouflage (a similar strategy has been adopted by some living decapods).

The similarity of *Mimetaster* to *Marrella*, the most common arthropod in the famous Middle Cambrian Burgess Shale of British Columbia (see Briggs *et al.* 1994), has long been noted. Recent analyses of arthropod relationships (Wills *et al.* 1995) confirm that *Mimetaster* forms a distinct group with *Marrella* and *Vachonisia*. They form a clade that is basal (primitive) with respect to all the other schizoramian arthropods (i.e. those with more than one branch on the limbs).

Vachonisia

Vachonisia is one of the rarest and more unusual arthropods in the Hunsrück Slate. It was redescribed by Stürmer and Bergström in 1976. Less than 10 specimens have been found, and only four of these have been prepared from the ventral side. The entire body was covered by a large carapace. The superficial similarity of this structure to the shell of a large brachiopod led to the temporary loss of one of the critical specimens among examples of a different phylum!

At the front of the carapace of *Vachonisia* there was a pronounced notch, and a clear median line is evident in some specimens. Although there are no examples that show the two halves as separate valves there may, nevertheless, have been a functional hinge. The central part of the carapace was strongly concave forming a heart-shaped depression, tapering rearward, which accommodated the body and trunk limbs. Eyes have not been identified. A pair of short, stout antennae at the front of the head was followed by a number of large segmented limbs increasing in size rearwards. Two undescribed specimens, which provide the best evidence of the ventral morphology, show that there were at least four pairs of these head limbs (Figs. 99, 100). They consisted of about six segments, and may have been equipped with basal spines that assisted in feeding. The mouth was floored by a large oval labrum. The trunk, which tapered to the rear of the carapace, bore a long series of much smaller limbs, possibly as many as 80 pairs.

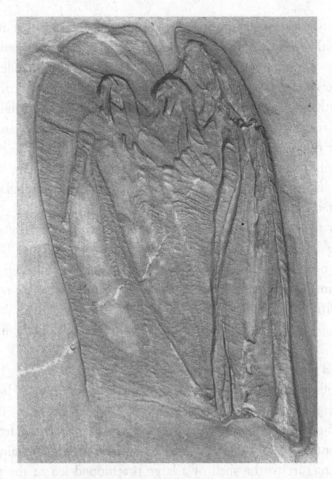

Figure 99 *Vachonisia rogeri* (Lehmann, 1955), ventral view, Eschenbach–Bocksberg mine, Bundenbach (× 1.4; HS 438).

Their structure is not well constrained, but they consisted of a segmented inner branch, and a filamentous outer branch (Fig. 101).

Vachonisia presumably walked on the head limbs, which are disproportionately larger than those of the trunk. Like *Mimetaster*, it was probably a deposit feeder, using the trunk limbs to stir up sediment in pursuit of small animals and organic particles. The shape of the carapace may have been adapted to influence feeding currents.

Vachonisia was originally considered to be a branchiopod crustacean (Lehmann 1955). Stürmer and Bergström (1976), however, noted the similarity between the appendages and those of *Mimetaster*, and of *Marrella* from the Cambrian Burgess Shale. Recent analyses of arthropod relationships (Wills *et al.* 1995) have confirmed that *Vachonisia* is most closely related to *Mimetaster* and *Marrella*. Together these arthropods form a clade that is basal (primitive) with

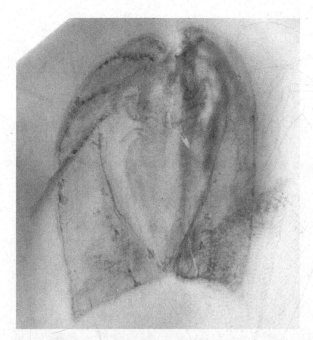

Figure 100 *Vachonisia rogeri*, (Lehmann, 1955), x-radiograph, ventral view, Bundenbach (× 1.0; Naturhistorisches Museum, Mainz: PWL 1994/53-LS, WB 300).

respect to all the other schizoramian arthropods (i.e. those with more than one branch on the limbs). New material of *Vachonisia* (see Figs. 99, 100), however, and recent developments in study techniques, suggest that further details of this remarkable Hunsrück Slate arthropod await discovery.

Crustaceans (subphylum Crustacea)

The Hunsrück Slate crustaceans are rare and are dominated by bivalved forms. The large phyllocarids like *Nahecaris* with preserved appendages contrast with the tiny poorly preserved ostracodes. These last have received little attention. The genera *Primitia, Zygobeyrichia, Carinakloedenia, Cornikloedenina,* and ?*Aparchites* have been recorded from interbedded coarser lithologies (Mittmeyer 1980*b*).

Nahecaris

Nahecaris (Figs. 102–104) is by far the most common of the six genera of phyllocarid crustaceans found in the Hunsrück Slate (Bergström *et al.* 1987) with specimens numbered in tens. There are at least three species of *Nahecaris,* but the majority of specimens are included in one, *N. stuertzi.*

Nahecaris, like other phyllocarid crustaceans, had a bivalved carapace

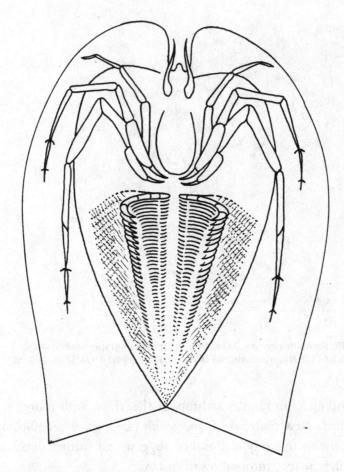

Figure 101 *Vachonisia rogeri* (Lehmann, 1955), restoration (from Stürmer and Bergström 1976).

covering the head, thorax, and anterior segments of the abdomen. It was redescribed by Bergström *et al.* (1987). A short anterior rostral plate, and longer posterior median plate, was inserted along the hinge line between the valves. There is some variability in both carapace shape and type of ornamentation even among specimens assigned to the species *Nahecaris stuertzi*, but these differences are at least partly preservational and cannot be used to separate individuals into distinct species. The head bore a pair of large eyes, and two pairs of biramous antennae, the second much larger than the first (Fig. 102). The mandible was large and robust, and is often preserved in strong relief, but the maxillae were small and are poorly known. The mouth was floored by a labrum.

There were eight thoracic segments, each bearing a similar pair of biramous limbs (the thoracopods) that decreased in size posteriorly. The segmented branch (endopod) had a pronounced 'knee' where the limb flexed downward. The ter-

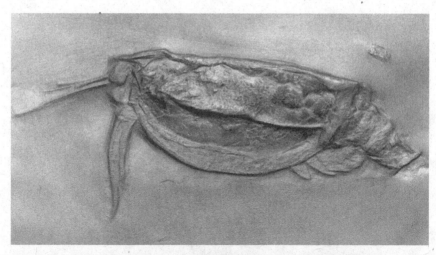

Figure 102 *Nahecaris stuertzi* Jaekel, 1921. Lateral view, distal part of the abdomen and furca missing, Eschenbach–Bocksberg mine, Bundenbach (×0.8; HS 322).

minal segment was fringed by setae. The exopod was in the form of a coarse comb with a row of large posteriorly directed finger-like projections. The abdomen consisted of seven cylindrical segments, longer than those of the thorax. The first five bore biramous limbs (the pleopods) made up of flap-like branches fringed with setae. The seventh abdominal segment was much longer than the others, and was followed by a short pointed telson flanked by a pair of much longer spines which formed the 'tail fork' (caudal furca) (Figs. 103, 104).

A second species was described and assigned to *Nahecaris* by Ferdinand Broili in 1930. It has a longer telson, with pronounced lateral spines. Only two specimens of *Nahecaris* ?*balssi* are known, however, and they are too poorly preserved to allow the genus to be identified with confidence (Bergström *et al.* 1989).

Specimens of *Nahecaris*, like *Mimetaster*, vary in their orientation to bedding. More are flattened dorsoventrally than laterally, however, suggesting that the valves were held widely apart in life. The unusual construction of the carapace hinge (which divides as a result of the insertion of plates along its length) may have limited the degree to which the valves could be brought together. *Nahecaris* is likely to have lived on the sea bed. It may have rested on the edges of the valves, supported by the antennae and the abdomen and tail, while using the small thoracic limbs for feeding. *Nahecaris* probably fed on carcasses as well as live animals. The large mandibles were equipped to deal with hard shells such as those of bivalve molluscs. The brush-like tips of the thoracic limbs may have prevented *Nahecaris* sinking into the muddy sediment; they may also have been used for

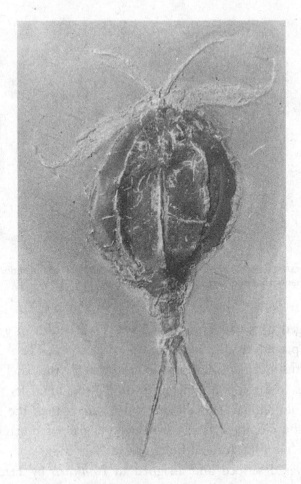

Figure 103 *Nahecaris stuertzi*, Jaekel, 1921. Dorsal view, specimen covered by water, dark areas indicate calcium phosphate, Eschenbach–Bocksberg mine, Bundenbach (×0.7; HS 546, photo Bartels).

cleaning the interior of the valves. Locomotion was mainly by swimming, using the abdominal limbs.

Nahecaris is a malacostracan, the group of crustaceans that includes the familiar lobsters and crabs. It belongs to the Phyllocarida, a diverse group of bivalved forms with less than 20 living species, grouped in the leptostracans. Due to the exceptional preservation of the Hunsrück Slate fossils, *Nahecaris* is one of the most completely known fossil phyllocarids. It belongs to the Paleozoic order Archaeostraca, which ranged from the Ordovician to the Permian.

Heroldina

Heroldina (Fig. 105) is one of the rarest arthropods found in the Hunsrück Slate; there are less than 10 known specimens (Bergström *et al.* 1989). It reached up

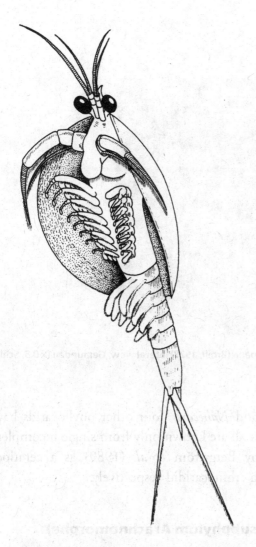

Figure 104 *Nahecaris stuertzi* Jaekel, 1921, restoration in ventrolateral view (from Bergström *et al.* 1987).

to 60 cm in total length, and is therefore the largest arthropod, and one of the largest animals, in the Hunsrück Slate. Some specimens preserve evidence of a small eye below the pronounced rostrum. The cuticle of the abdomen was more strongly ornamented than that of the carapace. The last abdominal segment was much longer than the others, and carried a very elongate telson. The limbs are poorly preserved and their morphology is unknown.

The mode of life of *Heroldina* may have been similar to that of *Nahecaris*. Like *Nahecaris* it belongs to the order Archaeostraca, but it lacks a median plate in the hinge, and therefore falls in a different suborder.

Figure 105 *Heroldina rhenana* (Broili, 1928) lateral view, Gemünden (×0.5, Schlosspark-Museum, Bad Kreuznach).

In addition to *Heroldina* and *Nahecaris*, four other phyllocarids have been found in the Hunsrück Slate but all are known only from single incomplete specimens. They were identified by Bergström *et al.* (1989) as a ceratiocarinid, ?*Dithyrocaris*, ?*Montecaris*, and a ?rhinocaridid respectively.

Arachnomorphs (subphylum Arachnomorpha)

Trilobites

The trilobites, which reached their acme in the Cambrian and Ordovician, declined in diversity in the Silurian and Devonian and were reduced to a single order in the Carboniferous, the Proetida, that finally became extinct toward the end of the Permian. The dominant order in the Devonian was the Phacopida and the most important Hunsrück Slate examples of this group are considered in the following pages. These pyritized trilobites are particularly important for the evidence that they provide for the morphology of appendages and internal organs. A number of other trilobites are known from the Hunsrück Slate, but they are very rare and often incompletely known. Soft tissues are not preserved. They include the phacopid genera *Comura* and *Treveropyge*, a number of homalonotids, the styginid *Scutellum* and the proetid *Cornuproetus* (these last two on the basis of a few specimens and a single fragment, respectively). Material of the odontopleurid *Ceratocephala* has also been recorded but has yet to be described.

Asteropyge

Although *Asteropyge* (Figs. 106, 107) is relatively common in the Hunsrück Slate the species-level systematics of the material has not been investigated, and it is possible that more than one species is represented (Stürmer and Bergström 1973). The cephalon projected slightly anteriorly and there were prominent genal spines. The wide glabella was flanked by large schizochroal eyes, but details of their internal structure are rarely preserved. Likewise few specimens preserve details of the limbs. In the head, a pair of antennae at the anterior was followed by three pairs of biramous limbs lying behind the mouth. The axial part of the head was

Figure 106 *Asteropyge* sp. Dorsal view, Eschenbach–Bocksberg mine, Bundenbach (× 1.5; SNG 029).

Figure 107 *Asteropyge* sp. Two partly enrolled individuals, one in dorsal and one in near lateral view, Eschenbach–Bocksberg mine, Bundenbach. (× 1.0; HS252).

occupied by the digestive glands and stomach, which led posteriorly into the intestine. The number of thoracic segments in individuals from the Hunsrück Slate varies from 10 to 12. The pleural spines were circular in cross section. The pygidium was also fringed with spines. The limbs of the thorax and pygidium, like those at the rear of the head, were biramous, but details of the morphology are unknown.

Asteropyge is assumed to have lived on the muddy bottom. It used the limbs in feeding, but in the absence of more data on their morphology, we can only speculate on how they functioned. *Asteropyge* probably fed on carcasses and small animals in the sediment. It could enrol for protection (Fig. 107).

Chotecops

Chotecops is the most abundant arthropod and the most common of the macro-fossils found in the Hunsrück Slate (Figs. 108–116). It is particularly important

Figure 108 *Chotecops* sp., dorsal view, Eschenbach–Bocksberg mine, Bundenbach (×1.3; HS 718).

as one of the few trilobites known that preserve details of the limbs. This familiar trilobite was first described in 1880 by Kayser who named it *Phacops ferdinandi* after Ferdinand Roemer, the author of the first accounts of fossils from the Hunsrück Slate published in 1863. The trilobite was recently investigated by Wolfgang Struve (Struve and Flick 1984, Struve 1985) who assigned it to a different genus, *Chotecops*. Struve (1985) described two new species and a number of other forms (in some cases subspecies) of *Chotecops* from the Hunsrück Slate based on a range of measurements, and on the pattern of sculpture on the glabella. A discussion of the differences between all these forms of *Chotecops* is beyond the scope of the present treatment, but it is likely that some of them simply reflect differences in preservation.

The cephalon of *Chotecops* was semicircular in outline. The large glabella expanded anteriorly, with tubercles arranged in various patterns on the surface

(Fig. 108). The head bore a pair of long antennae, consisting of more than 20 segments. There were three additional pairs of head appendages, each biramous like those of the trunk. The mouth was concealed by a triangular-shaped hypostome. The eyes of *Chotecops*, like those of *Asteropyge*, were schizochroal, made up of large individual lenses arranged in near vertical rows (Fig. 109) and separated by cuticle that is impermeable to light. These eyes are characteristic of phacopid trilobites which reached their maximum diversity in the Devonian. They contrast with the truly compound holochroal eyes (from which they probably arose) which comprise a much greater number of tiny polygonal lenses (up to 15 000), each in contact with its neighbour, the whole covered by a single cornea.

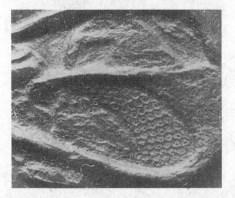

Figure 109 Schizochroal eye of *Chotecops* sp. showing the arrangement of the lenses (× 3.2; SNG 032).

Figure 110 *Chotecops* sp. Radiograph of the cephalon showing the stomach and adjacent diverticulae (digestive glands). (The stomach is evident in place in the specimen illustrated in Fig. 94) (× 6.0, WS 1605).

Holochroal eyes occurred in a much wider variety of trilobites than schizochroal eyes. Some trilobites lost the eyes and became blind.

Chotecops is remarkable in preserving traces of the digestive organs in the head region. These are revealed by x-rays (Stürmer and Bergström 1973). The radiating brush-like structures in Fig. 110 are thought to represent gut diverticula (or the 'liver' or hepatopancreas) that flanked the stomach.

The thorax of *Chotecops* usually consisted of 11 segments. The axis of the exoskeleton was defined by pronounced furrows flanked by a row of tubercles. The biramous limbs beneath are clearly displayed in Figs. 111 and 112, both of which provide a ventral view of specimens that are flattened slightly obliquely. The limbs consisted of an inner walking branch of seven segments, the last ringed by slender spines, and an outer gill branch with a fringe of filaments (Fig. 113) (Stürmer and Bergström 1973). Both these branches were attached to the coxa at the base of the limb, which articulated with the body. The segmented walking branches are most obvious, including the three pairs attached at the rear of

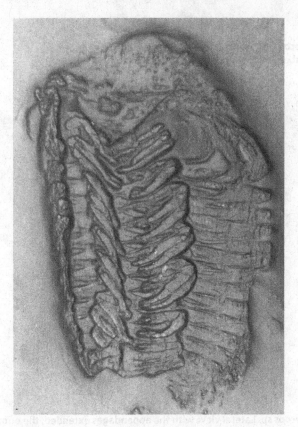

Figure 111 *Chotecops* sp. Ventral view showing appendages, Bundenbach (× 1.2; SNG 014).

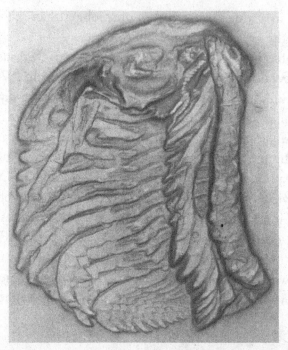

Figure 112 *Chotecops* sp. Ventrolateral view showing appendages, Eschenbach–Bocksberg mine, Bundenbach (× 1.1; private collection).

Figure 113 *Chotecops* sp. Lateral view with the appendages extended; the outer filamentous branches are exposed, particularly on the posterior appendages, Eschenbach–Bocksberg mine, Bundenbach (× 1.2; HS 214).

Figure 114 *Chotecops* sp. Enrolled, showing the pygidium and doublure of the cephalon. The hypostome has become separated from the doublure and is partly concealed by the pygidium. The inner surface of the right eye is evident between the margins of the cephalon and pygidium; the left eye, on the other hand, is seen from the external side. Bundenbach (× 1.2; Senckenberg, collection Brassel, SNG 023).

the cephalon. The gill branches are much more delicate and may be evident compacted against the dorsal exoskeleton. It is generally accepted that they functioned in respiration and perhaps in swimming.

Biramous limbs were also present beneath the pygidium. The axial region of the pygidium was clearly segmented, but traces of the segmentation became less obvious toward the margins. The rear margin of the pygidium was rounded and lacked spines.

Many Hunsrück Slate specimens of *Chotecops* are preserved in a partially enrolled state, with either the cephalon or pygidium tucked under the thorax. This may be a response to being overwhelmed by an influx of sediment. *Chotecops* displays a number of adaptations for enrolling. There was a vincular furrow around the ventral margin of the cephalon (Fig. 114) into which the margin of the pygidium inserted during complete enrolment. As in other trilobites, each thoracic tergite was equipped with an articulating half-ring in the axial region which passed beneath the tergite in front so that none of the vulnerable soft tissues was exposed when the trilobite enrolled (Fig. 115).

Moulting in *Chotecops* and closely related forms differed from that in most trilobites. The cephalon lacked functioning facial sutures and was therefore moulted in one piece. It became inverted as it separated from the thorax and pygidium, which remained essentially intact as the trilobite emerged forwards.

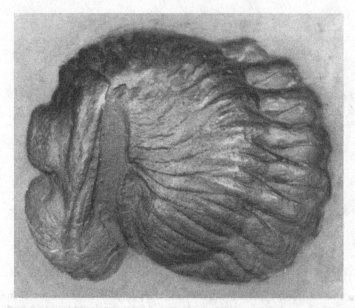

Figure 115 *Chotecops* sp., enrolled, lateral view, Kreuzberg mine, Weisel, Taunus, (× 1.3; SNG 026).

Freshly moulted exoskeletons of *Chotecops* display an arrangement (Fig. 116) characteristic of this type of moulting, which was termed Salterian (after J.W. Salter who first described it) by Rudolf Richter in 1937.

Chotecops was a benthic trilobite, walking on and swimming above the substrate. The rings of spines at the end of the walking limbs prevented them from sinking into the mud. Clear impressions of them are evident on some trace fossils (see Fig. 221). The appendages were also used for capturing animals. This food was processed by the strong coxae in the head region which functioned as gnathobases (jaws). The coxae of the thoracic limbs may also have assisted in feeding, by passing food forward toward the mouth along a ventral food groove. The gill branches may also have been used to set up feeding currents.

Odontochile

Odontochile is one of the rarer trilobites in the Hunsrück Slate. The largest individuals reached lengths of over 20 cm, but complete specimens are uncommon and only one that preserves traces of the appendages is known (Fig. 117).

The cephalon of *Odontochile* was flat and roughly semicircular in outline, with a wide flat border. Large schizochroal eyes flanked the glabella. The thorax comprised up to 12 segments. The long pygidium had 18 to 20 segments, with a narrow axis that extended posteriorly into a strong spine, and a broad flat area laterally.

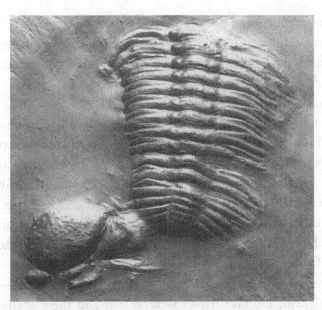

Figure 116 *Chotecops* sp., exuviae (moulted exoskeleton) in the Salterian configuration, Bundenbach (× 1.2; SNG 038).

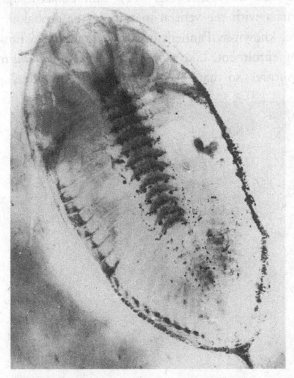

Figure 117 *Odontochile rhenanus* (Kayser, 1880), Bundenbach. (× 0.6; private collection, Stürmer x-radiograph.)

Odontochile was presumably a bottom-dwelling trilobite. The low relief on the exoskeleton may have facilitated its concealment in the sediment. In the absence of evidence of the limb morphology, the details of its mode of life are unknown.

Parahomalonotus

Parahomalonotus is unusual in that, even though it is one of the more abundant trilobites from the Hunsrück Slate, no specimens are known that preserve any of the limbs or other soft tissues (Brassel and Bergström 1978). This may reflect the rarity of complete specimens at the classic localities around Bundenbach and Gemünden. The genus *Parahomalonotus* occurs more frequently in the slates of the Eifel and Taunus regions, where the sediment includes a higher proportion of silt, than in the Hunsrück region. The Hunsrück Slate species, *P. planus*, has not been recorded from elsewhere.

The cephalon of *Parahomalonotus* was short with a curved anterior margin. The eyes were very small. Dorsal furrows were absent and the glabella was not very pronounced. The thorax comprised 13 segments. The pygidium was very short and also rounded. The morphology of the cephalic margin and pygidium was adapted to allow spiroidal enrolment (Fig. 118) – the dorsal margin of the pygidium came into contact with the ventral margin of the cephalon. A small tubercle on the doublure, known as Pander's organ, functioned in limiting the amount of overlap during enrolment. Like *Chotecops*, *Parahomalonotus* may have lacked functional facial sutures, so that the cephalon was moulted as an entire

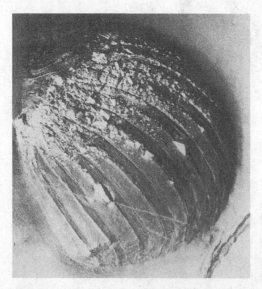

Figure 118 *Parahomalonotus planus* (Koch, 1863). Enrolled, the surface of the specimen covered in large euhedral crystals of pyrite, Katzenberg mine, Mayen (× 1.1; HS265).

unit. In the moult illustrated in Fig. 119 the cephalon lies at the posterior margin of the pygidium, dorsal side up.

In the absence of preserved appendages we can deduce little detail of the likely mode of life of *Parahomalonotus*. The generally smooth, effaced exoskeleton of homalonotids (Fig. 120) has been interpreted as an indication that they were burrowing forms, normally living below the sediment surface (Bergström 1973). This would appear to make them more susceptible to burial and more likely to be preserved intact than other trilobites, but nevertheless pyritized soft tissues are unknown. Complete specimens may simply have died and decayed in their burrows, where they were protected from disarticulation. Rapid influxes of sediment may not have completely filled the burrow, allowing decay to proceed

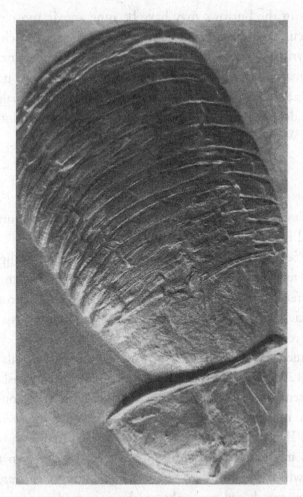

Figure 119 *Parahomalonotus planus* (Koch, 1863). Exuviae (moulted exoskeleton) in the Salterian configuration, Kreuzberg mine, Weisel, Taunus (× 1.0; private collection).

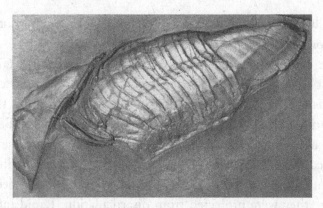

Figure 120 *Parahomalonotus planus*, (Koch, 1863). Dorsolateral view, the cephalon separated from the rest of the exoskeleton, Katzenberg mine, Mayen (× 0.6; HS 431).

unhindered. Conditions in the burrows may have become anoxic too rapidly to allow pyritization to occur. Analyses of sediments yielding the Hunsrück fossils, however, suggest that pyritization of soft tissues only occurred where concentrations of iron were anomalously high (Briggs *et al.* 1996). This may not have applied in the sediments preferred by homalonotids. Alternatively homalonotids may not have burrowed deeply, lying instead in shallow excavations in loose sediment on the sea floor (Whittington 1993).

Rhenops

Three species of *Rhenops* are known from the Hunsrück Slate: *R. limbatus*, *R. anserinus* and *R. lethaeae*. This genus is important in that a small number of specimens preserve good evidence of the morphology of the limbs.

The cephalon of *Rhenops* was semicircular in outline, projecting slightly in the mid- line. There were stout genal spines. The glabella expanded anteriorly, and had pronounced lateral furrows. It was flanked by large schizochroal eyes. There were eleven thoracic segments. The pygidium had a well-defined axis, tapering posteriorly, and the rear margin was fringed by stout spines.

At least two specimens of *Rhenops* are known that preserve details of the appendages (Figs. 121, 122). The difference in the outline of their exoskeletons suggests that they represent different species, but their limbs may be considered together. The specimen illustrated in Fig. 121 was discovered first, and has received more attention. Bergström and Brassel (1984) argued that it shows four pairs of biramous limbs in the head, in addition to the antennae, even though most other trilobites in which the limbs are preserved, including *Chotecops*, show only three. This assertion was based not only on the preserved position of the limbs in the cephalon, but also on the identification of 15 pairs (interpreted as

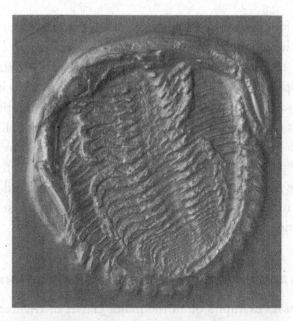

Figure 121 *Rhenops* cf. *anserinus*. Ventral view showing appendages, original of Bergström and Brassel (1984), Eschenbach–Bocksberg mine, Bundenbach (× 1.2; SNG 028).

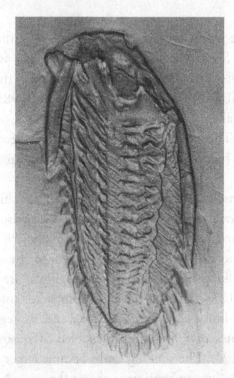

Figure 122 *Rhenops* sp. (Koch, 1863). Ventral view with appendages, Eschenbach–Bocksberg mine, Bundenbach (× 1.0; HS261).

4 cephalic and 11 thoracic) anterior of an apparent change in limb morphology marking the anterior of the pygidium. This interpretation is equivocal, however, because the cephalon of this specimen was clearly distorted and foreshortened during compaction (the specimen is almost circular in outline), and it is not clear where the boundary between the limbs of the thorax and those of the pygidium lies. The second specimen of *Rhenops* (Fig. 122) is much less distorted. Here there is no obvious distinction between the limbs of the thoracic and those of the pygidium, and just three pairs of biramous limbs appear to belong to the head.

Rhenops, like *Asteropyge* and *Chotecops*, was a bottom-dwelling trilobite. It probably used the limbs to feed on carcasses and small animals in the sediment. It could enrol for protection.

Chelicerates and their allies (superclass Cheliceriformes)

The Hunsrück Slate includes examples of all the major classes of cheliceriformes with aquatic representatives, the Xiphosura, Eurypterida and Scorpionida, as well as the Pycnogonida (sea spiders).

Weinbergina

The xiphosurans, which include the recent horse-shoe crabs, are one of the rarest arthropod groups in the Hunsrück Slate. The only genus, *Weinbergina*, is represented by just six specimens, two of which have been discovered since the most recent detailed description was published by Stürmer and Bergström (1981). The fifth specimen (Fig. 123) is the smallest known at 7.5 cm long, and reveals the morphology of the ventral side. The sixth, held in a private collection in Bundenbach, shows the dorsal side but does not add significantly to our knowledge of this arthropod.

The body was divided into a prosoma covered by a large anterior shield, an opisthosoma consisting of ten separate segments, and a tail spine or telson. The exoskeleton was strongly convex dorsally (Fig. 124). The prosoma bore a pair of small chelicerae, which were used to cut up prey and push it into the mouth, followed by six similar pairs of walking limbs. These walking limbs are clearly evident on the new specimen (Fig. 123), those on the right compacted under the edge of the prosoma. The walking limbs consisted of a coxa and six segments, and terminated in an array of spines that presumably enabled *Weinbergina* to walk on the muddy substrate (Fig. 124). The walking limbs became larger toward the rear of the prosoma. None of the known specimens shows the eyes. A closely related genus, *Legrandella* from South America, had a pair of lateral compound

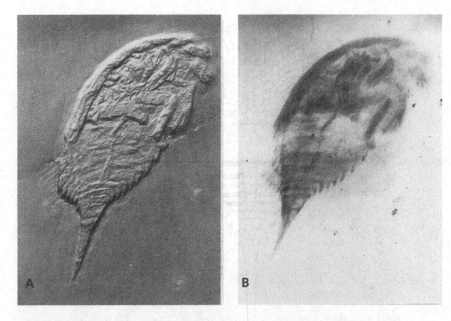

Figure 123 *Weinbergina opitzi* Richter and Richter, 1929. The smallest specimen, showing the ventral morphology including the limbs, Eschenbach–Bocksberg mine, Bundenbach: (A) surface; (B) x-radiograph. (× 1.0; HS 328, WS 12867).

eyes and a median eye; *Weinbergina* may have been similar (Stürmer and Bergström 1981).

The opisthosoma of *Weinbergina* consisted of 10 articulated segments. The tergites of the first seven carried three rows of nodes, one in the mid-line, the other two flanking it on each side. The anterior six opisthosomal segments bore gill lamellae ventrally, equivalent to those in *Limulus*. The gill lamellae in each of these limbs were concealed by a flat plate of cuticle that was fringed along its straight posterior margin by a row of spines. Traces of these are evident in Fig. 123 which also shows the course of the gut. The posteriormost three segments of the opisthosoma were enclosed by rings of cuticle. The telson was triangular in cross-section.

Weinbergina is assumed to have lived mainly on the muddy sea bottom. The morphology of the prosomal limbs is rather generalized (see Fig. 124). They no doubt functioned in walking, the distal spines providing purchase on the soft mud. A possible xiphosuran trace fossil was reported by Richter (1941, p. 250), but the wide and irregular spacing of the imprints flanking the median groove leave its identity uncertain. Although the prosomal limbs were not specifically adapted for burrowing, *Weinbergina* was no doubt capable of digging into soft sediment. It probably fed on small animals such as worms which it caught on or near the sediment surface. The chelicerae were used to cut up prey and transfer

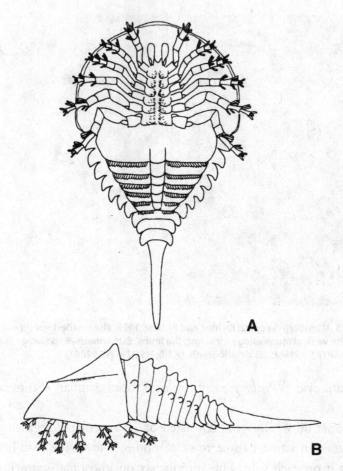

Figure 124 *Weinbergina opitzi* Richter and Richter, 1929, restoration: (A) dorsal; (B) lateral (from Stürmer and Bergström 1981).

it to the mouth. *Weinbergina* was probably able to swim on its back using the prosomal and opisthosomal limbs to generate thrust in a manner similar to modern *Limulus*.

Weinbergina is unique among xiphosurids in having six walking limbs (in addition to the chelicerae) in the prosoma in contrast to the usual complement of five. The sixth pair is assumed to be equivalent to the chilaria of *Limulus*, small plates which form rudimentary limbs at the rear of the prosoma. *Weinbergina* belongs within the Synziphosurina which is an extinct sister group to the Limulina, the suborder that includes the living horse-shoe crabs.

Rhenopterus

Eurypterids (sea scorpions) are one of the rarest animals in the Hunsrück Slate (Kutscher 1975*a*). They were large predatory arthropods that originated in marine environments in the early Ordovician but colonized fresh water in the late Paleozoic before becoming extinct at the end of the Permian. Eurypterids were formerly grouped with the xiphosurids (including *Weinbergina*) in a mainly aquatic group called the Merostomata of equivalent rank to the mainly terrestrial Arachnida. This twofold division of chelicerates has now been shown to be artificial and the eurypterids are considered a separate class. The eurypterids are characterized by large sizes. About half the families had representatives over 80 cm long. A smaller number, particularly the pterygotids and their relatives, reached huge dimensions of up to 180 cm.

Only a single fragmentary eurypterid specimen has been reported from the Hunsrück Slate. It was identified by Lehmann (1956*b*) as *Rhenopterus diensti*, which had previously been described (Størmer 1939) on the basis of much more complete material from the older Klerfer Schichten at Willwerath in the Eifel (Fig. 125). The comments below are based mainly on this material. *Rhenopterus* was a small eurypterid, about 5 cm in length.

Eurypterids are characterized by a division of the body into a prosoma and opisthosoma, followed by a telson. The prosoma bore lateral eyes and six pairs of limbs. The first pair, the chelicerae, were used for pushing food into the mouth. The other prosomal limbs in *Rhenopterus* were long and segmented, and terminated in spines. The anterior six segments of the opisthosoma (the mesosoma) carried gills ventrally. The following segments (of which only one is known) lacked limbs.

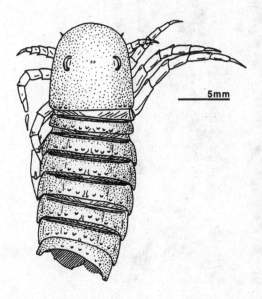

Figure 125 *Rhenopterus diensti* Størmer, 1939, dorsal restoration (after Størmer 1939).

The morphology of the limbs of *Rhenopterus* suggest that it walked on the substrate. Although it lacked the swimming paddle characteristic of some eurypterids, it was probably also capable of swimming. *Rhenopterus* presumably preyed on small invertebrates although it may also have been a scavenger. Featureless fragments of large arthropods from the Hunsrück Slate of the Taunus region may represent eurypterids.

Palaeoscorpius

The position of the scorpions within the chelicerates is uncertain, and it is preferable to assign them to a separate class. Like the eurypterids, scorpions are represented in the Hunsrück Slate by just one specimen (Fig. 126). This large individual is 12.5 cm long. It was discovered by Lehmann (1944) in a concretion within a piece of roof slate. An x-radiograph allowed the specimen to be identified and facilitated its preparation to expose both dorsal and ventral sides on the upper and lower surfaces of the piece of slate (Kutscher 1971*b*). Certain features were impossible to reveal and are only known from the evidence of the x-radiograph.

Palaeoscorpius displays the familiar morphological features of the scorpions. The prosoma bore six pairs of appendages. The first, the small chelicera, is not

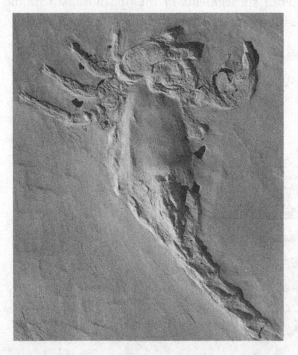

Figure 126 *Palaeoscorpius devonicus* Lehmann, 1944, Bundenbach (×0.75).

evident on the specimen. The pedipalps, which follow, ended in large powerful claws (chelae). Posteriorly there were four pairs of segmented walking limbs each with three terminal spines. The prosoma was followed by a broad segmented mesosoma, the segments of which bore broad abdominal plates ventrally. The trunk tapered into the presumably flexible metasoma. The telson is not preserved.

Modern scorpions are exclusively terrestrial and the mode of life of fossil forms was long assumed to have been similar. Silurian examples, however, occur in marine or marginal marine sediments. It is now generally accepted that early scorpions were aquatic. Respiratory organs are rarely preserved in fossil scorpions but gill-like structures are evident in *Waeringoscorpio hefteri* from the Lower Devonian of Alken (see Selden and Jeram 1989). It is likely that the abdominal plates of *Palaeoscorpius* also concealed gills and this, together with its association with marine organisms, suggests that it was probably marine. The large pedipalps confirm that, like other scorpions, it was a predator.

Cheloniellon

Cheloniellon is one of the largest and rarest arthropods in the Hunsrück Slate. Only five specimens are known. Nevertheless Wilhelm Stürmer's x-radiograph of one of them (Fig. 127) has provided perhaps the most striking and familiar image of any of the Hunsrück Slate fossils. This specimen, and two others, were the only specimens of *Cheloniellon* known when Stürmer and Bergström (1978) redescribed this arthropod. However, they recognized that a specimen of the supposed myriapod, *Bundenbachiellus giganteus*, was likely to be *Cheloniellon*. A new, well-preserved specimen, incomplete in the anterior part of the head, was discovered in 1991 (Fig. 128). It is considerably smaller than the others and apparently represents a juvenile. This specimen is prepared from the ventral side and lacks the long posterior caudal furca.

The body outline of *Cheloniellon* was oval. The specimens are flattened in different orientations to the bedding, and the small number of examples makes it difficult to determine the length:width ratio of the outline with confidence. The restoration (Fig. 129) is based on Stürmer and Bergström's (1978) estimate of 1.3:1.0. These drawings also incorporate their interpretation of how the limbs corresponded to the tergites, a relationship that was distorted in the specimens due to decay and compaction.

Twelve articulating tergites formed the dorsal shield. The first two covered the head tagma, the following eight the trunk, and the last two, reduced considerably in size, the abdomen. A pair of large compound eyes was situated on the dorsal surface of the first tergite; the lenses are not preserved. Ventrally

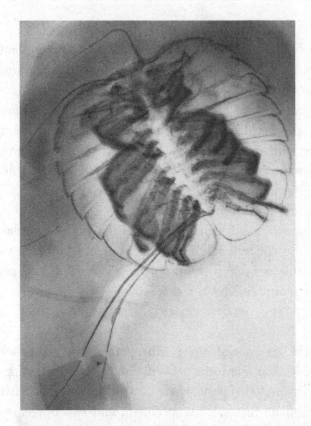

Figure 127 *Cheloniellon calmani* Broili, 1932, Bundenbach, radiograph, ventral view
(×0.5, Schlosspark-Museum, Bad Kreuznach, KGM 1983/269; WS 2487).

the head tagma bore a pair of slender antennae anteriorly followed by five pairs of uniramous segmented limbs. The first of these lay in front of the mouth and was equipped with bundles of setae at its base; it may have been sensory. The remaining four formed a series of feeding appendages with well-developed inwardly-facing spiny gnathobases. These were used to crush and comminute the food through movements of the limbs. The last of these feeding limbs appears to have been attached to the second tergite.

The trunk bore eight pairs of biramous limbs, each corresponding in position to a dorsal tergite. The inner segmented branch lacked a gnathobase. It terminated distally in a group of tiny spines. The outer branch consisted of a broad flat lobe with an array of strong filaments around the margin, particularly posteriorly. It presumably functioned as a gill and probably also in swimming.

The last two trunk segments of *Cheloniellon* were surrounded by lateral extensions of the tergites in front of them, and they differed from the rest. The ninth carried a pair of elongate furcal rami, longer than the rest of the body, that

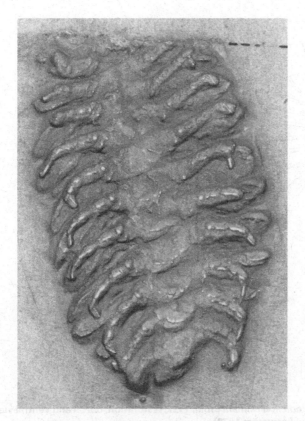

Figure 128 *Cheloniellon* sp. Juvenile specimen, ventral view, part of the head and the furcal rami have been lost, Eschenbach–Bocksberg mine, Bundenbach (× 2.0; HS 525).

was apparently attached dorsally. The terminal segment was a conical telson-like structure that lacked appendages.

Cheloniellon was a benthic animal. It used the segmented branch of the trunk limbs to walk on the muddy substrate and was probably also able to swim. The strong spiny gnathobases on the head appendages indicate that it was a carnivore, probably preying on soft-bodied animals in a manner similar to modern horse-shoe crabs.

Cheloniellon falls within the arachnomorphs, a large group of arthropods including the chelicerates and trilobites. Analyses of its relationships show that it occupies a position as a sister taxon to the chelicerates (Wills *et al.* 1995).

Sea spiders (class Pycnogonida)

The position of the pycnogonids (sea spiders) within the Arthropoda has been a subject of controversy, but they are generally regarded as closer to the chelicerates than to any other group. There are over 1000 known living species in over 80 genera. The Hunsrück Slate pycnogonids are very important as the only known

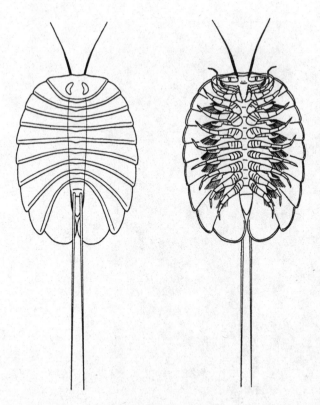

Figure 129 *Cheloniellon calmani* Broili, 1932, restoration, dorsal and ventral views (from Stürmer and Bergström 1978).

fossil examples, apart from a possible larval form described by Müller and Walossek in 1986 from the Upper Cambrian of Sweden, and undescribed pyritized species from the Middle Jurassic near La Voulte-sur-Rhône in France. The pycnogonids of the Hunsrück Slate were redescribed by Bergström *et al.* in 1980. The body of sea spiders consists of a cephalosoma with four pairs of appendages, a thorax with three pairs, and a segmented abdomen. The limbs of the cephalosoma are (1) the cheliciphore, which is used in cutting up prey, (2) the palp which is sensory and may function in manipulating food, (3) the oviger, which occurs only in the male and is specialized for carrying the eggs until the larvae develop, and (4) the first locomotory limb, which is used in both walking and swimming. There are small eye tubercles on the dorsal side of the cephalosoma, and a snout-like proboscis on the ventral side, with the mouth at its distal extremity. Each of the three thoracic segments bears a pair of walking legs, usually consisting of nine podomeres and terminating in a claw. Thus the pygnogonids have a total of four pairs of walking limbs. The abdomen, which lacks limbs, consists of five segments and a telson.

Palaeoisopus

Palaeoisopus is is one of the most remarkable animals of the Hunsrück Slate (Figs. 130–132). It is perhaps not surprising that in the original description, published in 1932, Broili interpreted the abdomen as a proboscis, and the chelicerae as a paddle-shaped abdomen, thus orienting the body back-to-front. *Palaeoisopus* is the most abundant of the Hunsrück Slate pycnogonids as well as the largest, reaching a limb span of up to 40 cm. It has only been found in the Bundenbach area.

The cephalosoma bore a large eye tubercle dorsally. The proboscis is often curved beneath the cephalosoma where it is not evident from the dorsal side. The large chelifores projected anteriorly, the terminal chelae often overlapping. The palps and ovigers are similar and it is often difficult to distinguish one from the other. The large walking limbs appear to have been laterally flattened in life. They consisted of eight segments, and terminated in a pronounced claw. X-radiographs of the first of these limbs have revealed what appear to be pyritized muscles in some specimens (Bergström *et al.* 1980). The limb segments were armed with an array of spines on the ventral margin. Articulation between them

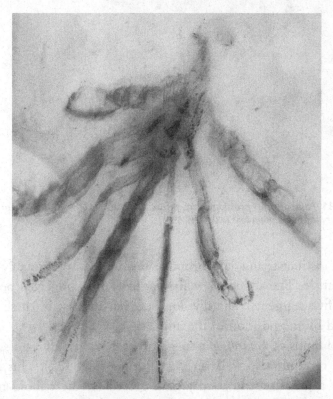

Figure 130 *Palaeoisopus problematicus* Broili, 1928, radiograph (× 0.6; WS 12446).

Figure 131 *Palaeoisopus problematicus* Broili, 1928. Dorsal view. The first right ambulatory limb has been lost (× 0.5; HS 456).

was facilitated by distinct triangular arthrodial membranes, presumably covered by thin flexible cuticle. The abdomen, which lacked limbs, projected posteriorly and consisted of five segments. The division between the last of these and the telson was marked by the position of the anus.

The flattened limbs of *Palaeoisopus* suggest that it was an efficient swimmer, although it could certainly also walk on the substrate (Fig. 132). This large sea spider was clearly a predator and it may have grazed on the crinoid meadows (Bergström *et al.* 1980). The claws could have been used to grasp the stems or

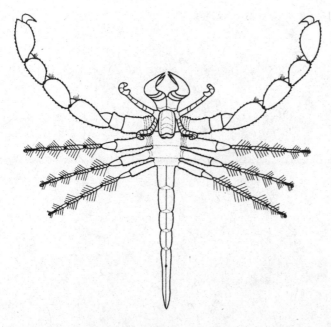

Figure 132 *Palaeoisopus problematicus* Broili, 1928, restoration of the ventral side (after Bergström *et al.* 1980).

arms of crinoids and pull them within reach of the chelifores. This interpretation is suggested by the association of some specimens of *Palaeoisopus* with crinoids, and by the occurrence of crinoids with severed arms. Cut-up fragments of starfish are sometimes preserved in the Hunsrück Slate, suggesting that they too may have been part of its diet.

Palaeopantopus

The small pycnogonid *Palaeopantopus* is one of the rarest animals in the Hunsrück Slate. The term 'pantopod' means 'only appendages' and refers to the outline of the organism which appears to consist only of limbs. The original description was based on just two specimens (Broili 1928*b*). A third was found by Wilhelm Stürmer while seeking material for x-ray investigation in the collections of the Humboldt Museum in Berlin (Bergström *et al.* 1980). The fourth was discovered and prepared by Christoph Bartels (Fig. 133).

At first glance the most striking difference between *Palaeoisopus* and *Palaeopantopus* is the size. The limb span of *Palaeopantopus* does not exceed 10 cm whereas in *Palaeoisopus* it reaches up to 40 cm. The body of *Palaeopantopus* was flattened. There were slender palps and ovigers, and small chelifores are evident on the new specimen (Fig. 133). The larger limbs of *Palaeopantopus* were

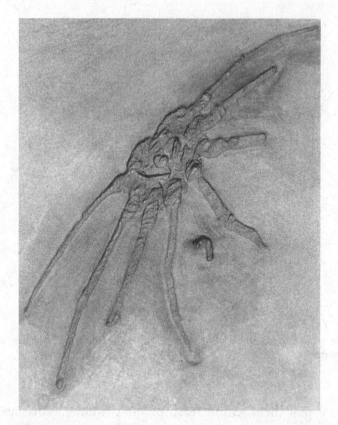

Figure 133 *Palaeopantopus maucheri* Broili, 1929. Ventral view (× 1.25; HS 437).

cylindrical in cross-section and slender, tapering distally to a point. There was a short dorsally-directed abdomen.

The morphology of the limbs suggests that *Palaeopantopus* moved slowly on the sediment surface, the limbs holding the body well above the substrate. The absence of any clear adaptation for swimming does not preclude this mode of locomotion, although *Palaeopantopus* was clearly a less effective swimmer than *Palaeoisopus*. In the absence of detailed evidence for the nature of the chelifores and proboscis the nature of feeding cannot be deduced.

Palaeoisopus and *Palaeopantopus* are thought to be relatively primitive in morphology compared with *Palaeothea*, the third Hunsrück Slate pycnogonid, which is the only one that can be assigned to an extant order.

Palaeothea

Only one specimen of this tiny Hunsrück Slate animal, which does not exceed 1 cm in dimension, is known (Fig. 134A). It was discovered by Stürmer while x-raying slabs of slate, and the matrix has not been removed to reveal it (Bergström

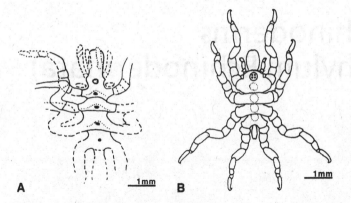

Figure 134 *Palaeothea devonica* Bergström *et al.*, 1980. Bundenbach: (A) interpretative drawing of the radiograph (WS 2287); (B) drawing of the extant pantopod *Pigrogromitus timsanus* Calman. (From Bergström *et al.* 1980.)

et al. 1980). Thus we could argue that it is a taxon that has never actually been seen! Nevertheless the morphology is sufficiently well preserved to reveal that it alone among Hunsrück Slate pycnogonids can be assigned to a living order, the pantopods.

There was a series of tubercles along the dorsal axis of *Palaeothea*, the most anterior of which was presumably an eye tubercle. The cylindrical proboscis projected forwards. The appendages of the cephalosoma are poorly preserved apart from the large first walking limb. The thorax bears three additional, presumably similar, walking limbs. The abdomen extended posteriorly beyond this. In spite of the small size of *Palaeothea*, its morphology indicates that it is unlikely to represent a larval form of either of the other two Hunsrück Slate pycnogonids, but rather that it is a separate taxon (Bergström *et al.* 1980). The habitat of *Palaeothea* was clearly benthic. It presumably moved slowly along the sea bed with a more hanging stance than *Palaeopantopus*, keeping the body close to the sediment surface.

Comparisons are hampered by the limited morphological information preserved, but Bergström *et al.* (1980) considered *Palaeothea* to have been very similar to modern pantopods (Fig. 134B). Among the Hunsrück Slate pycnogonids, its relationships lie closer to *Palaeopantopus* than to *Palaeoisopus*.

7 Echinoderms (phylum Echinodermata)

General remarks

The first illustrations of Hunsrück Slate fossils, published by Ferdinand Roemer in 1862, showed echinoderms – crinoids and starfishes (see Fig. 1). Crinoids and starfishes (asteroids and ophiuroids) remain the most familiar and abundant of the Hunsrück Slate fossils, as well as being some of the most spectacular (Figs. 135, 155). Many of the genera and species, particularly of the starfishes, are

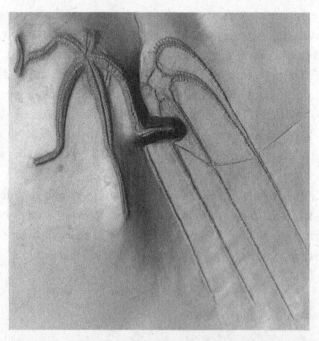

Figure 135 *Urasterella asperula* Roemer, 1863 (on the left) and *Furcaster palaeozoicus* Stürtz, Eschenbach–Bocksberg mine, Bundenbach (× 0.6; HS 144).

unique. About 30% of the crinoid and more than 80% of the starfish species recorded from the Palaeozoic of the Rhenish Massif are known only from the Bundenbach region.

The Echinodermata is one of the most distinctive of all invertebrate phyla. Echinoderms ('spiny skinned animals'), which are exclusively marine, have a skeleton of calcite plates. These plates are mesodermal in origin, and are covered by a thin outer skin. The plates may be independent or fused together to form a rigid skeleton. Many echinoderms bear spines, which articulate on tubercles on the plates. Spines are most obviously developed in the sea urchins (Echinoidea). The phylum is characterized by a fivefold radial symmetry, although a bilateral symmetry is superimposed upon this in a number of echinoderm groups.

Perhaps the most diagnostic feature of echinoderms is their water vascular system, a network of internal canals which gives rise to tube feet that extend beyond the body. Water enters the system through the madreporite, a perforated plate on the dorsal surface. It is transported down the vertical stone canal to the ring canal which surrounds the mouth. Radial branches from this ring canal give rise to the tube feet which project through the skeleton of echinoids and run the length of the arms of starfishes and crinoids. The position of these radial canals, normally five in number, and their associated tube feet define the ambulacral areas of the echinoderm and its skeleton. The tube feet, which are controlled by muscular sacs, perform a variety of functions including respiration, locomotion and feeding.

Some types of echinoderms have a good fossil record, because of the high preservation potential of their calcite skeletons. In most circumstances, however, the skeleton disarticulates rapidly after death, particularly where the plates are not fused. Thus intact asteroids and ophiuroids are very rare in the fossil record. The Hunsrück Slate echinoderms are remarkable for their completeness. Rapid sedimentation and burial on the Hunsrück Slate sea floor prevented the plates from becoming separated (Fig. 136). Pyritization strengthened the skeletons and often preserved the soft tissues. Many starfishes, for example, preserve not only the skeleton and spines, but also the skin, revealing the outline where it extends between the arms. Crinoids, asteroids and ophiuroids are abundant in parts of the Hunsrück Slate, but the other echinoderm taxa, including the edrioasteroids, echinoids, and holothuroids, are very rare.

The classification scheme followed here is that used in Benton (1993), based on cladistic analyses of all the major groups by Smith (1984) and Paul and Smith (1984). They identified a major division into two subphyla: the Pelmatozoa, stemmed forms including the crinoids, cystoids and blastoids, and the

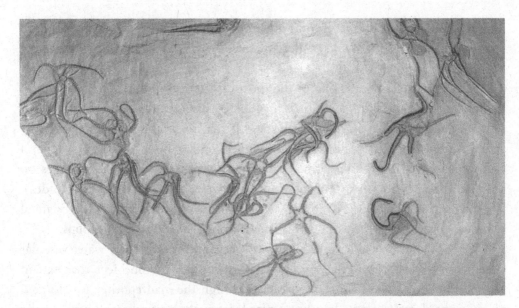

Figure 136 A large group of *Furcaster palaeozoicus* Stürtz, 1886, overwhelmed by a rapid influx of sediment, Eschenbach–Bocksberg mine, Bundenbach (× 0.5; HS 459).

Eleutherozoa, including the edrioasteroids, asteroids, ophiuroids, echinoids and holothuroids, which lack a stem. The asteroids and ophiuroids constitute distinct classes and are not united in a group Asterozoa. However, we use the informal term 'starfishes' here where it is convenient to refer to the asteroids and ophiuroids together. The phylogeny of these various groups is far from resolved in detail.

Homalozoans

The homalozoans are the most unusual echinoderms. The body was dorsoventrally flattened and shows no sign of a radial or pentameral symmetry; indeed the symmetry is superficially bilateral. Although the homolazoans are treated as echinoderms here, there is an alternative view that they are ancestral chordates (Jefferies 1986, 1990, 1997: see Benton 1993, Chapter 26 for a summary of Jefferies' views on the classification of these forms). This radical idea, which results in conflicting interpretations of both their orientation and the nature of various morphological features, echoes the evidence of embryology and molecular sequence data for a close relationship between echinoderms and chordates. Regardless of whether the homalozoans are considered echinoderms or ancestral chordates, a consensus appears to be emerging that they form a natural clade. On that basis it is reasonable to infer that the stem (aulacophore in mitrates, stele in solutes) is a homologous organ in all homalozoan groups (C.R.C. Paul, personal communication).

Three species of homalozoans have been described from the Hunsrück Slate: the mitrates (class Stylophora) *Mitrocystites styloideus* and *Rhenocystis latipeduncu-lata* (Figs. 137, 138), and the solute (class Homoiostelea) *Dehmicystis globulus* (Fig. 139). They have not been studied since the original work of Dehm (1932, 1934). Homalozoans in general, however, have received considerable attention (Caster 1967, Ubaghs 1967*a,b*, Jefferies 1968, 1984, 1986) and it is clear that the examples from the Hunsrück Slate require revision (*Rhenocystis* is currently being investigated under the direction of R.P.S. Jefferies). A number of additional species, including that illustrated in Fig. 140, await description.

The mitrates are characterized by a tendency toward a bilateral arrangement of plates. The flattened body (theca) was surrounded by marginal plates which enclosed smaller plates on the dorsal surface and larger irregular plates on the ventral. The marginal plates were perforated by several pores connected to the water vascular system. The mitrates are characterized by a robust stem, which here is assumed to define the posterior margin. The anus was positioned at the base of this stem and the mouth at the other end of the theca (these openings are interpreted in the opposite sense by some echinoderm workers, who consider the aulacophore to represent an arm). Solutes are similar to the mitrates in their asymmetrical flattened theca. They differ, however, in having an anterior arm (brachiole).

Mitrates (class Stylophora)
Mitrocystites
Mitrocystites styloideus was described by Dehm (1934) on the basis of a single poorly preserved specimen and doubtfully assigned to this genus. Some 20 additional specimens are now known. Fig. 137, which illustrates the ventral surface, shows the robust median spines that project from the plates of the stem.

Rhenocystis
Rhenocystis (Fig. 138) is relatively common at Bundenbach but has not been found at other Hunsrück Slate localities. The theca is roughly rectangular, and the animal is characterized by two long spines that project from the extremities of the anterior margin. The elongate stem could presumably lever the body off the substrate, and also functioned in locomotion. Recently discovered trace fossil evidence shows how regular lateral movements of the stem of *Rhenocystis* propelled the animal along the surface of the sediment.

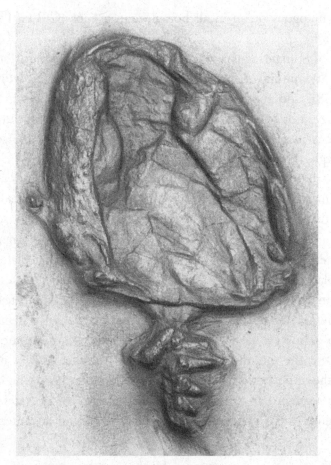

Figure 137 *Mitrocystites styloideus* Dehm, 1934, ventral view, Eschenbach–Bocksberg mine, Bundenbach (× 3.0; HS 287).

Solutes (class Homoiostelea)

Dehmicystis

Dehmicystis was originally described on the basis of an incomplete specimen (Dehm 1934) which does not preserve the arm. An additional specimen has since been found (Fig. 139). The theca was approximately circular in outline. Caster (1967) noted an anal pyramid on Dehm's original specimen close to the base of the stem. The stem was broad proximally and extended into a long flexible distal extremity. It was presumably used in locomotion, like that in mitrates.

Pelmatozoans (subphylum Pelmatozoa), except crinoids

The pelmatozoans include the various forms of stemmed echinoderms, most of which were attached to the substrate, at least as adults. The crinoids were by far

Figure 138 *Rhenocystis latipedunculata* Dehm, 1932, dorsal view, Eschenbach–Bocksberg mine, Bundenbach (× 2.0; HS296).

the most diverse and abundant pelmatozoans in the Hunsrück Slate sea. There were also rare blastoids and rhombiferan cystoids. Crinoids are common in other marine Devonian sequences, but specimens are usually disarticulated. The importance of the Hunsrück Slate pelmatozoans lies in their remarkably complete preservation.

Blastoids (class Blastoidea)

Blastoids were stemmed echinoderms characterized by a highly developed pentameral symmetry. The petal-shaped calyx consisted of 13 or 14 fused basal, radial

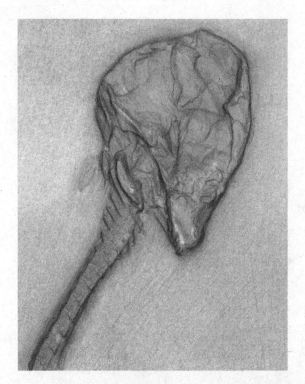

Figure 139 *Dehmicystis globulus* Dehm, 1934, Eschenbach–Bocksberg mine, Bundenbach (× 1.4; HS 320).

and deltoid plates. A large number of slender unbranched brachioles arose from biserial ambulacra. In some blastoids these were confined to the oral (upper) surface, in others they extended down the calyx to the stem. There were five openings (the so-called hydrospires) in the roof of the calyx surrounding the mouth. These hydrospires were connected to the ambulacral system. The calyx was supported by a slender stem consisting of smooth columnals (Fig. 141). The earliest blastoids are known from the Ordovician, they reached a peak in the Carboniferous, and died out in the Permian. Two species of blastoids are known from the Hunsrück Slate (Kutscher 1965*c*).

Pentremitidea

Pentremitidea is the larger and more robust of the two Hunsrück Slate blastoids. The brachioles were confined to the upper surface of the calyx. The original description (Jaekel 1895) of the Hunsrück species, *P. medusa*, was based on four specimens on a single slab from Kaub on the Rhine (Schmidt 1934, pl. 23; Kutscher 1965*c*, pl. 7) which has subsequently disappeared. A few additional specimens have been found at Bundenbach (Fig. 142).

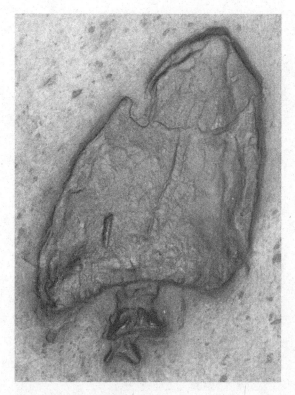

Figure 140 Unidentified homalozoan gen. et sp. indet., Eschenbach–Bocksberg mine, Bundenbach (× 2.0; HS).

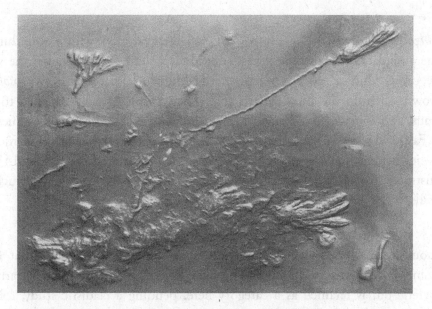

Figure 141 Blastoidea undetermined, Bundenbach (× 0.75; SNG 073).

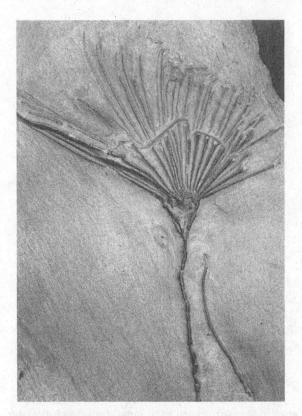

Figure 142 *Pentremitidea medusa* Jaekel, 1895, Eschenbach–Bocksberg mine, Bundenbach (× 1.3; HS 453).

Schizotremites

Schizotremites (Figs. 143, 144) had an elongate oval calyx with a long thin stem. The ambulacra, which bore long slender brachioles, extended down the sides of the calyx. Only a few specimens of the Hunsrück Slate species *S. osoleae* are known, all from the Bundenbach area. They were originally assigned to a new genus, *Pentremitella*, by Lehmann (1949) but the generic identity was questioned by Fay (1967, S417; see Kutscher 1967*b*), who suggested that the Hunsrück Slate species belongs to *Schizotremites*. The specimen illustrated here (Figs. 143, 144) illustrates how x-radiographs can assist in the preparation of Hunsrück Slate fossils.

Rhombiferan cystoids (class Rhombifera)

Rhombiferans are characterized by rows of respiratory openings that form a rhomb-shaped pattern on adjoining plates. The class Rhombifera is clearly poly-phyletic but is retained as a category here, pending a cladistic study, following the arrangement in Benton (1993).

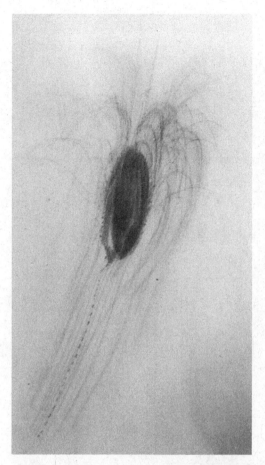

Figure 143 *Schizotremites osoleae* (Lehmann, 1949), radiograph, Eschenbach–Bocksberg mine, Bundenbach (× 1.4, HS 323; WS 12 854).

Regulaecystis

Regulaecystis pleurocystoides (Figs. 145, 146), the sole cystoid known from the Hunsrück Slate, has been found only in the Bundenbach area. The flattened theca was flask-shaped in outline. The delicate plates were strengthened by ridges. A ridge encircled the stem where it inserted into the basal plates. The large periproct (anal field), which was covered by small scale-like plates, was positioned in the centre of one side of the flattened theca (the anal: Fig. 145), allowing it to be distinguished from the other (the antanal: Fig. 146). A ridge surrounded the periproct, and the single rhomb. Dehm (1932) illustrated a specimen showing three additional stout ridges radiating from the centre of antanal side (see Kesling 1967, S197) but these ridges are not always evident (Fig. 145) and they may be a preservational artefact. Two robust arms projected from the tapering end of the

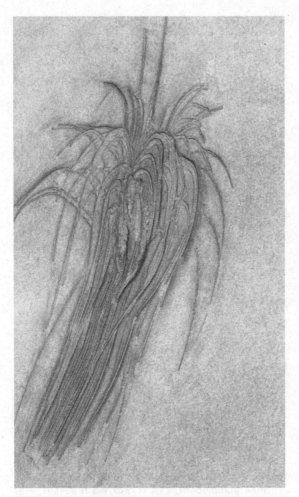

Figure 144 *Schizotremites osoleae* (Lehmann, 1949). Surface view of specimen in Fig. 143 (× 1.5, HS 323).

theca. The stem consisted of 20 to 30 simple stout proximal columnals followed by 60 to 80 more elongate barrel-shaped columnals distally.

The flattened shape of *Regulaecystis* suggests that it lay on the sediment surface. The stem might have served to hold on to objects on the sediment surface, allowing the body to lift off the substrate in a current. However, it is more likely that the mode of life was similar to that of the closely related Ordovician cystoid *Pleurocystites*. Paul (1967) argued that *Pleurocystites* lay on the sediment on the anal side, and moved forward using undulating movements of the stem. This benthic mode of life is similar to that interpreted for the Hunsrück Slate homalozoans which display a similar, but convergently evolved, morphology.

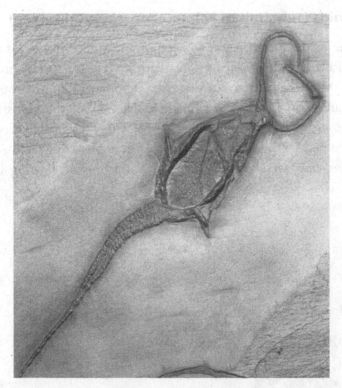

Figure 145 *Regulaecystis pleurocystoides* Dehm, 1932, anal view, Eschenbach–Bocksberg mine, Bundenbach (× 0.9; HS 345).

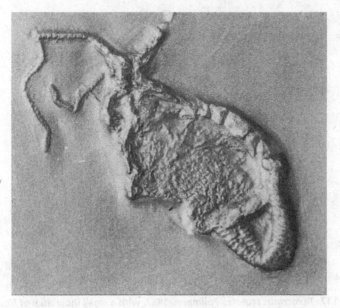

Figure 146 *Regulaecystis pleurocystoides* Dehm, 1932, antanal view, Bundenbach (× 1.4; SNG 115).

Crinoids (subphylum Pelmatozoa, class Crinoidea)

Crinoids are one of the most diverse and important groups in the fossil record, particularly in Paleozoic limestones. More than 5500 fossil species are known, approximately 700 of them represented in the Devonian. More than 60 species and subspecies are known from the Hunsrück Slate. All of the Hunsrück Slate crinoids were sessile, at least as adults (Fig. 147), with the possible exception of *Senariocrinus* (see Fig. 159). Other fossil crinoids were planktonic and some, like the Jurassic *Seirocrinus*, were pseudoplanktonic, attaching to floating logs. The

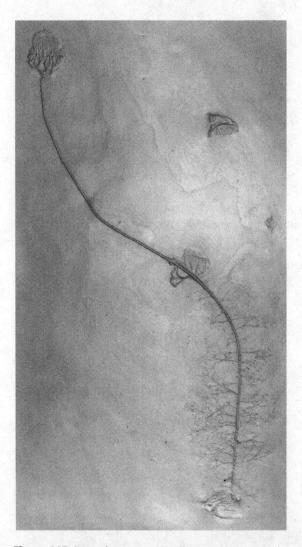

Figure 147 *Taxocrinus stuertzii* Follmann, 1887, with a small individual of *Hapalocrinus* sp. attached to the stem. An unidentified rugose coral is also present on the slab. Eschenbach–Bocksberg mine, Bundenbach (×0.3; HS 133).

body of crinoids consists of a cup with a flexible or rigid roof (tegmen), together called the crown. The cup bears the arms (brachia), and is supported by the stem. Crinoid stems rarely exceed a metre in length, and lengths of this order are known only in the taxocrinids among the Hunsrück Slate forms. The stem plates or columnals take a variety of shapes but are usually round, oval or pentangular in outline. The stems of some crinoids give rise to cirri that support the animal above the sediment surface and/or act as anchors (Fig. 147). The mouth is situated in the tegmen, which is therefore referred to as ventral (as it is equivalent to the ventral surface of starfishes). The arms carry the tube feet, which are used in feeding.

The arrangement of plates in the cup of crinoids is an important character in identifying taxa. The base of the cup may be formed by a single circlet of plates, the basals (monocyclic), or a second circlet, the infrabasals, may insert beneath these (dicyclic). The basal plates support radial plates which in turn bear the arms. In some cases the proximal arm plates (brachials) are incorporated into a calyx. The arms consist of one or two rows of brachial plates (i.e. they are uniserial or biserial). The anus is situated on the upper surface, between two of the arms, and is often surrounded by additional plates. It may be elevated above the calyx on an anal tube or an anal sac (see Fig. 162, 166) which releases the faeces away from the arms.

Crinoids feed by turning the dorsal side of the crown to face the current and spreading the arms wide. The arms are usually branched, and these branches may give rise to pinnules. The whole forms an extensive filtering apparatus. As the water current passes through the arms and pinnules, food particles and small organisms are captured on the tube feet. This food is transported down the arms to the mouth. This type of feeding, using the lift provided by the current, is favoured by most stalked crinoids. Sediment stirred up by water currents is detrimental to crinoids because of the risk of particles clogging the water vascular system or food grooves. The arms and stem of most Hunsrück Slate crinoids are long and slender (see Figs. 160, 165). Recent crinoids that live in environments where currents are prevalent but not sufficiently strong to disturb the sediment are similar in construction.

The classification of crinoids has changed radically since Schmidt (1934, 1941) monographed over 40 species and subspecies from the Hunsrück Slate. The most recent classification (Simms and Sevastopulo 1993) recognizes three subclasses, Camerata, Disparida and Cladida (the status of a fourth, the Hybocrinida, is uncertain). This scheme acknowledges the polyphyletic nature of the Inadunata (see Benton 1993) and abandons it, elevating the constituent taxa,

Disparida and Cladida, as subclasses. The Flexibilia and Articulata are included as infraclasses within the Cladida.

Sixty-five species and subspecies of crinoid, now assigned to 30 genera, have been described from the Hunrück Slate (Follmann 1887, Jaekel 1895, Haarmann 1922, Klähn 1929, Koenigswald 1930a,b, Opitz 1932, Schmidt 1934, 1941, Lehmann 1955, Kutscher 1966c, 1970b, 1976a, Kutscher and Sieverts-Doreck 1973). Advances in our understanding of fossil crinoids, together with the wealth of new material now available, makes a revision of the Hunsrück Slate crinoids overdue. Here we illustrate and briefly describe representatives of 18 genera.

Crinoids are common at many localities in the Hunsrück Slate outside the classic Bundenbach area. They are particularly abundant in the south-eastern Eifel region where more sandy facies dominate. Although they are usually disarticulated, these crinoids can frequently be identified. Some specimens from these areas have been figured here (see Figs. 18, 148–150), but they have yet to be studied by a specialist.

Camerates (subclass Camerata)

The camerates appear to be a natural group, although relationships within them are problematic. They are characterized by a calyx of fused plates. The mouth is situated beneath the tegmen plates on the surface of the calyx, and the anus may open through these plates or from an anal tube. The proximal parts of the arms are usually incorporated into the calyx where they are separated by interradial plates.

Acanthocrinus

One of the most striking fossils ever discovered in the Hunsrück Slate was the holotype of *Acanthocrinus rex* (Jaekel 1895, Schmidt 1934; see Ubaghs 1978, Fig. 232), one of the rarest of Hunsrück Slate crinoid species, which was lost during the second world war.

The calyx of *Acanthocrinus* (Fig. 148) was large and bowl shaped with a slight depression where the stem was attached. Long spines projected from the rows of plates flanking the stem, and other plates bore tubercles, protecting the calyx from predators. Each arm divided three times near its base, which was incorporated into the calyx, resulting in a total of 40 free branches. The arms were biserial and bore densely spaced pinnules. The proximal elements of the pinnules were elongate and slender, the distal ones shorter. The robust stem consisted of large ridged plates and shorter intervening ones. There were no cirri. The distal end of the stem was typically enrolled in a spiral fashion that presumably helped

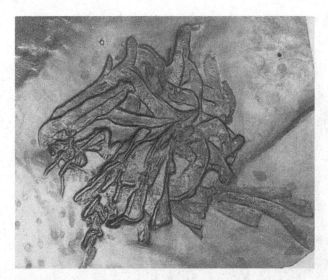

Figure 148 *Acanthocrinus rex* Jaekel, 1895. Note the presence of *Chondrites* in the lower left, Kreuzberg mine, Weisel, Taunus (× 0.5; private collection).

to anchor it in the sediment. Three species, *A. rex*, *A. heroldi* and *A. lingen-bachensis*, are known from the Hunsrück Slate.

Culicocrinus

Culicocrinus is one of the rarest Hunsrück Slate fossils. The calyx consisted of three unequal-sized basal plates and five large radial plates, and incorporated the first two plates of each arm. Each radial plate bore two downwardly-projecting slender spines. The arms branched twice, each producing four free arms per ray. They were biserial, with a long slender pinnule on every brachial. Only one species of *Culicocrinus*, *C. spinatus* (Fig. 149), occurs in the Hunsrück Slate itself. Several species are known from the more sandy Lower Emsian of the Rhenisches Schiefergebirge.

Diamenocrinus

Diamenocrinus stellatus was originally described from the Lower Devonian Rhenish Normal Facies. It is very rare in the Hunsrück Slate; the first specimen was found by Lehmann in 1955. A second specimen, from the Wispertal area of the Taunus, is illustrated here (Fig. 150). The calyx was globose. The plates display a characteristic star-shaped pattern of ridges, from which the species name *stellatus* is derived. The proximal parts of the arms were incorporated into the calyx where they formed pronounced ridges. The arms divided four to five times giving a total of between 60 and 90 free arms. Much of the distal part of the

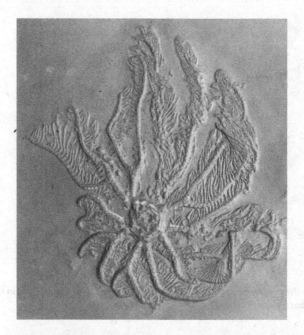

Figure 149 *Culicocrinus spinatus* Jaekel, 1895, Kreuzberg mine, Weisel, Taunus (×0.8; SNG 082).

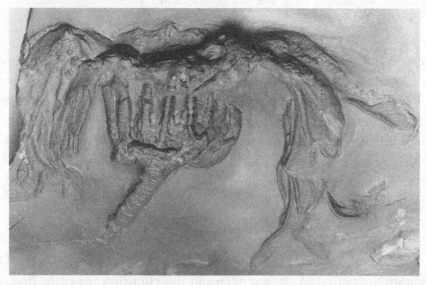

Figure 150 *Diamenocrinus stellatus* Jaekel, 1895, Nordberg mine, Sauerthal, Taunus (×0.6; HS 464).

arms has been lost in the illustrated specimen, revealing the calyx. The pentagonal stem is a diagnostic feature of this crinoid. A second species, *D. opitzi*, also occurs in the Hunsrück Slate (Schmidt 1934, pl. 33).

Hapalocrinus

Schmidt (1934, 1941) described several species and subspecies of *Hapalocrinus*. These taxa are difficult to discriminate, however, and some of the variation may be preservational.

The calyx of *Hapalocrinus* was made of three unequal basal plates and five large radial plates. The arms were uniserial and branched at least once. Proximally they bore pinnules on every second brachial. Beyond the seventh of these, however, a pinnule was attached to every brachial. The stem was slender and consisted of relatively long columnals.

Hapalocrinus elegans (Figs. 151–153) had 10 slender free arms, and the calyx was characterized by narrow grooves flanked by parallel ridges. Since it was described by Schmidt (1934) as rare, the Eschenbach–Bocksberg quarry at Bundenbach has yielded several well-preserved examples, some consisting of bush-like groups of up to 50 individuals at all stages of development.

Fig. 154 illustrates the characteristic habit of *Hapalocrinus frechi*. Individuals of this species usually have 20 free arms as a result of branching at least twice. Spines projected from the dorsal surface of these arms and also from the radial plates, beneath the insertion of the arms. The stem was round, the individual elements tapering in both directions. *H. frechi* is very similar to *Imitatocrinus* in general appearance, but the contruction of the calyx is very different.

Hapalocrinus innoxius (Fig. 155) differs from the other species in its long

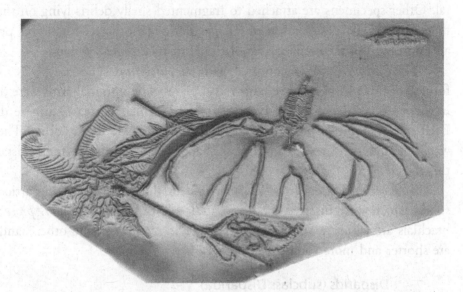

Figure 151 *Hapalocrinus elegans* Jaekel, 1895 (on the left), *Parisangulocrinus zeaeformnus* Follmann, 1887 (on the right), and the asteroid *Eospondylus primigenius* (Stürtz, 1886) (at the bottom) Bundenbach (×0.8; SNG 086).

Figure 152 *Hapalocrinus elegans* Jaekel, 1895, Eschenbach–Bocksberg mine, Bundenbach (× 0.9; SNG 105).

arms and pinnules. Spines were lacking on both the arms and calyx. The stem was relatively short and consisted of short round elements.

Thallocrinus

Thallocrinus (Fig. 156) is relatively common in the Hunsrück Slate of the Bundenbach region. The slab illustrated in Fig. 157 shows both adult and several juvenile specimens, the smallest of which is less than 1 cm in total length. The extremity of the stem of one of the adults is coiled into an anchoring spiral. Other specimens are attached to fragmented shelly debris lying on the sediment surface. This association of *Thallocrinus* with other echinoderms provides an impression of a colonized area of the Hunsrück Slate sea floor.

Thallocrinus is easily confused with *Hapalocrinus*. The calyx was likewise formed of three basal and five large radial plates. The uniserial arms were divided into two equal branches (see Fig. 156). The brachials decreased in size distally. Each bore a pinnule, projecting alternately to one side or the other. The stem was long and slender with hour-glass shaped elements in the more distal two-thirds.

Thallocrinus differs from *Hapalocrinus* in the presence of tiny articulating spines which were inserted on the ventral side of the arms. In *T. procerus* the brachials are elongate and slender. In *T. hauchecornei*, on the other hand, they are shorter and more robust.

Disparids (subclass Disparida)

The disparids were formerly considered to be an order of the Inadunata. With the recognition that the inadunates are polyphyletic, the Disparida was elevated

Figure 153 Group of six *Hapalocrinus elegans* Jaekel, 1895, one *Parisangulocrinus* sp. (in the centre between the stems) and the asteroid *Taeniaster beneckei* (Stürtz, 1886) near the base of the stems, Eschenbach–Bocksberg mine, Bundenbach (×0.6; HS 129).

to the status of subclass (Simms and Sevastopulo 1993). The plates of the calyx of these crinoids were usually fused but the proximal elements of the arms were not incorporated into it. The disparids were monocyclic (the cup consisted of one circlet of plates).

Figure 154 *Hapalocrinus frechi* Jaekel, 1895, ssp. indet. Fifteen specimens ranging in size and age, Eschenbach–Bocksberg mine, Bundenbach (× 0.8; HS 375).

Calycanthocrinus

Calycanthocrinus (Fig. 158) is common in the Hunsrück Slate, particularly around Gemünden. Due to its fancied resemblance to a broomstick this crinoid is commonly referred to as the 'Gemünden broomstick' ('Gemündener Besen'). The basal plates of the cup were reduced from five to three. The five radials were augmented by four extra plates, forming a series of three larger and six smaller elements. Each of these nine plates bore a simply constructed arm consisting of elements that were two to three times longer than wide.

Figure 155 *Hapalocrinus innoxius* W.E. Schmidt, 1934 (*ca* 20 individuals), *Triacrinus koenigswaldi* W.E. Schmidt, 1934 (bottom left) and two examples of the asteroid *Eospondylus primigenius* (Stürtz, 1886) (left), Eschenbach–Bocksberg mine, Bundenbach (×0.3; HS 509).

Senariocrinus

The crown of *Senariocrinus* (Fig. 159) was bilaterally symmetrical. It bore five free arms without ramules or pinnules, and a stout anal sac that was about twice their thickness. The five free arms resulted from a twofold division of two, plus one that was unbranched. The anal sac was covered dorsally by 10 to 16 elements similar to those of the arms, and ventrally by a much larger number of round flat calcareous elements. It terminated in a small anal pyramid. The stem was very short and thin and consisted of only a small number of elements.

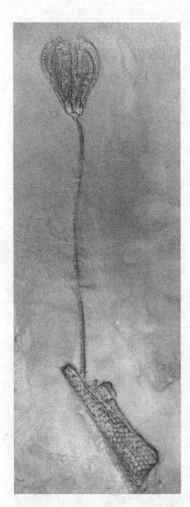

Figure 156 *Thallocrinus procerus* W.E. Schmidt, 1934, attached to an orthocone cephalopod, the shell of which is covered with an epizoan tabulate coral, Eschenbach–Bocksberg mine, Bundenbach (× 0.7; HS 442).

The family Calceocrinidae, to which *Senariocrinus* belongs, was unusual in having a highly derived morphology, which presumably indicates adaptation to a particular mode of life. The short thin stem of *Senariocrinus* tapered at its termination, and it seems unlikely that the crinoid was permanently attached to the substrate. The stem does not appear strong enough to have supported the calyx above the sediment surface. Schmidt (1934) suggested that *Senariocrinus* may have been capable of swimming, but its mode of life remains uncertain.

Triacrinus

Triacrinus is very common at Gemünden. It is characterized by a small conical cup and very long unbranched arms. The calyx consisted of three unequal basal

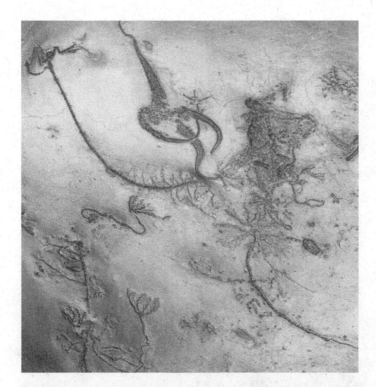

Figure 157 An assemblage of echinoderms buried *in situ*: several specimens of *Thallocrinus hauchecornei* Jaekel, 1895 (lower half of the photo), a *Triacrinus* sp. (upper left corner), a *Taeniaster beneckei* (Stürtz, 1886) (an unusual individual with only four arms), lying below a small *Ophiurina lymani* Stürtz, 1889 (centre, upper part of photo), Untereschenbach mine, Bundenbach (×0.5; HS 128).

plates, five radials and an anal. The fusion of some of the basals and radials to form larger composite plates resulted in a triangular cross-section of the calyx. The arms were simply constructed of elongate elements and reached lengths up to 20 times the height of the cup. There were no pinnules. There was an arm-like anal sac. *Triacrinus elongatus* (Fig. 160) is characterized by a small high cup with very elongate arms. *T. koenigswaldi* (Figs. 155, 161) is distinguished by thicker skeletal elements and a shorter truncated cup bearing shorter arms.

Subclass Cladida

Like the disparids, the cladids were formerly considered to be an order of the Inadunata. Simms and Sevastopulo (1993) elevated them to a subclass in which they included the infraclasses Cyathocrinina, Flexibilia and the post-Paleozoic Articulata. The cladids were dicyclic (the cup consisted of two circlets of plates). Most of the Hunsrück Slate representatives are stem-group cladids. A few can be assigned to the Cyathocrinina or Flexibilia.

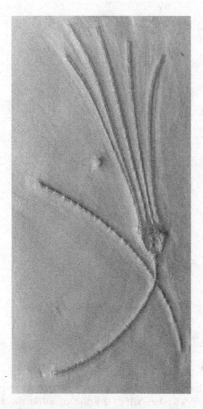

Figure 158 *Calycanthocrinus decadactylus* Follmann, 1887, Kaisergrube mine, Gemünden (× 1.0; SNG 101).

Stem group cladids

Bactrocrinites

The cup of *Bactrocrinites* (Fig. 162) was tall and narrow and this is reflected in the shape of the five infrabasals and five basals. The radials accommodated the arms in broad horse-shoe shaped insertions. There was an anal plate in the posterior. The arms were long and slender and divided three or four times. There were no pinnules. The anal sac was slender and longer than the arms, separating the faeces and the feeding current. The appearance of the stem is reminiscent of a string of pearls. Only the proximal part consisted of short pentangular columnals. Only one species from the Hunsrück Slate, *Bactrocrinites jaekeli*, is reliably assigned to this genus.

Bathericrinus

Bathericrinus is one of the rarer crinoids in the Hunsrück Slate. The robust cup was cone shaped and dicyclic. Two anal plates were inserted in the circle of five

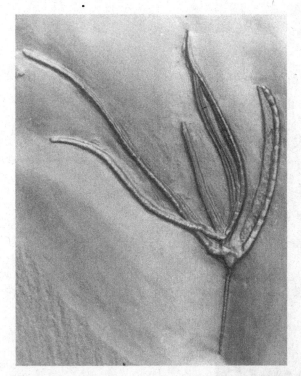

Figure 159 *Senariocrinus maucheri* W.E. Schmidt, 1934, Eschenbach–Bocksberg mine, Bundenbach (× 1.0, Private collection).

radials. The arms divided into two main branches, which gave rise to alternating lateral branches. Distally they bore robust dorsal spines. The most characteristic feature of this genus is the anal tube that bore 10 long spines at its termination (Fig. 163).

In addition to *Bathericrinus hystrix* (Fig. 163), *B. ericius*, *B. semipinnulatus* and *B. spaciosus* occur in the Hunsrück Slate. Kutscher (1968*a*) mistakenly assigned them to *Dictenocrinus* instead of *Bathericrinus*, and this has been perpetuated in some subsequent literature (Mittmeyer 1980*b*).

Dicirrocrinus

When Schmidt described *Dicirrocrinus* in 1934, only one incomplete specimen provided evidence of the calyx and arms. This remains the case, even though examples of the stem of this crinoid are relatively common. The cup was dicyclic and, according to Schmidt (1934), spherical in outline. The radial circlet included three intervening anal plates that supported a large anal sac. The morphology of the stem is very characteristic (see Fig. 26), allowing fragments to be identified easily. It bore many lateral cirri that branched several times forming a tree-like

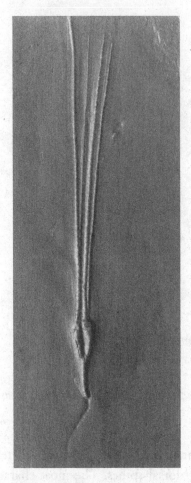

Figure 160 *Triacrinus elongatus* Follmann, 1887, Kaisergrube mine, Gemünden (× 1.2; SNG 102).

structure that extended over a considerable area of sediment. The stem cirri anchored the crinoid and may have supported it above the sediment surface.

Follicrinus

Follicrinus (Fig. 164) was one of the largest crinoids in the Hunsrück Slate with arms over 20 cm in length. The cup was very small and in the shape of a truncated cone. It was dicyclic and included five radials and two intervening anal plates. The arms were very slender and divided three or four times. Pinnules were absent. The stem was relatively short and tapered distally. *Follicrinus* had a large balloon-shaped anal sac covered with plates with a star-like ornament. *F. grebei* is relatively common at a single horizon in the Eschenbach–Bocksberg quarry but is rare elsewhere. A second species, *F. kayseri*, also occurs in the Hunsrück Slate.

Figure 161 *Triacrinus koenigswaldi* W.E. Schmidt, 1934 (on the left), *Imitatocrinus gracilior* (F. Roemer, 1863), and *Taxocrinus* sp. (on the right) associated with a partly decalcified rugose coral, Eschenbach–Bocksberg mine, Bundenbach (× 0.4; HS 311).

Gastrocrinus

Gastrocrinus (see Fig. 21) is relatively rare in the Hunsrück Slate. The cup was small relative to the size of the crinoid. It was dicyclic and included three anal plates in the circle of radials. The arms divided up to 14 times. The branching points were much more closely spaced distally than proximally resulting in bush-like terminations. *Gastrocrinus* is characterized by a very large long anal sac consisting of robust plates with thick transverse ridges. Long slender cirri arose in groups of five at intervals along the stem, projecting toward the cup. *Gastrocrinus* tends to be more prevalent in siltier horizons, suggesting that this crinoid may have favoured higher energy, sandier substrates. It is known from both the Bundenbach, Mayen, and Middle Rhine areas.

Imitatocrinus

Schmidt (1934) named this crinoid *Imitatocrinus* because of its general similarity to *Hapalocrinus*. The broad cup was dicyclic, with five radials and an intervening anal plate. Both the cup and the proximal part of the stem bore spines (these latter have not been reported previously). The arms were uniserial, and zigzagged distally where they too bore strong spines. Their arms are characterized by a large number of long slender minor branches (ramules). *Imitatocrinus*

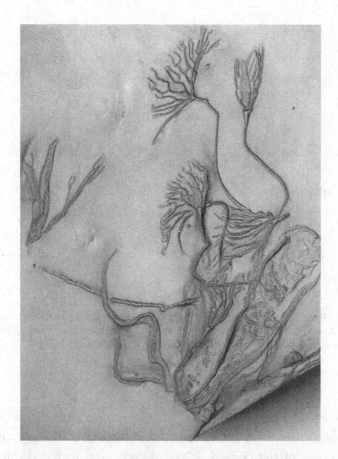

Figure 162 A group of six *Bactrocrinites jaekeli* (W.E. Schmidt, 1934) anchored to a shell (right half of photo), a *Parisangulocrinus zeaeformis* (Follmann, 1887) (on the left: note the long anal tube) and an individual of the very rare asteroid *Protasteracanthion primus* Stürtz, 1886 (at the bottom), which has become caught up in the crinoids and was presumably carried in with the sediment that buried them. Eschenbach–Bocksberg mine, Bundenbach (× 0.7; HS 122).

(Figs. 161, 165, 172) is relatively common in the Bundenbach area and is also known from the Taunus region (Mittmeyer 1980*b*).

Parisangulocrinus

Parisangulocrinus is relatively common in the Hunsrück Slate where it usually occurs in clusters or 'meadows'. The conical cup consisted of small infrabasals, large basals, and radial plates in circlets of five, the last with three intervening anal plates. The arms were slender, with broad ventral grooves covered by tile-shaped plates. They divided three or four times in variable positions resulting in 10 free arms per radius, giving a total of 50. There was a large, elongate anal sac. The stem was slender with pentangular proximal elements.

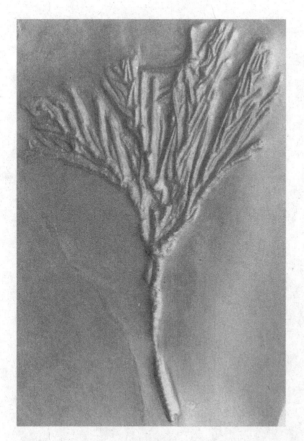

Figure 163 *Bathericrinus hystrix* (W.E. Schmidt, 1934), Eschenbach–Bocksberg mine, Bundenbach (× 1.2; SNG 085).

Four species of *Parisangulocrinus* are known from the Hunsrück Slate, two of which are illustrated here. The calyx of *P. minax* was broad, with vaulted plates that bore axial ridges. The anal tube was very elongate and characterized by long distal spines (right-hand specimen in Fig. 166). The calyx of *P. zeaeformis* (Fig. 167) was narrow and elongate. The anal sac was similar in appearance to a cob of maize, long and curved, with eight rows of plates.

Rhadinocrinus

The cup of *Rhadinocrinus* (Fig. 168) was relatively small and in the shape of a truncated cone. It consisted of small infrabasals, large basals, and radials in circlets of five, the last with three intervening anal plates. Each of the plates was slightly broader than high. The arms divided proximally into two main branches. The pattern of subsequent branching varied. In *R. dactylus* each division was separated by two lateral branches, the total number of branches varying with the

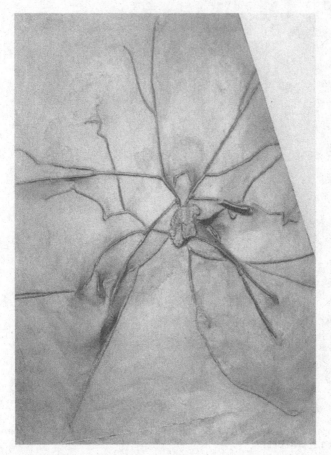

Figure 164 *Follicrinus grebei* Follmann, 1887, Eschenbach–Bocksberg mine, Bundenbach (× 0.5; HS 398).

size of the individual. The anal sac was very long and slender and was usually curved. In many individuals it is hidden among the numerous branches of the arms, and is only evident on x-radiographs.

Rhenocrinus

The cup of *Rhenocrinus* (Fig. 169) was small and cone shaped. It was dicyclic, and incorporated four anal plates that supported a large elongate cylindrical anal sac. This sac is characterized by clearly defined vertical rows of plates. The long slender arms did not branch, but bore a large number of long ramules on alternate sides. *Rhenocrinus ramosissimus* is relatively common in a restricted number of localities, such as the Rosengarten roof-slate mine near Bundenbach, suggesting that it had very specific habitat requirements.

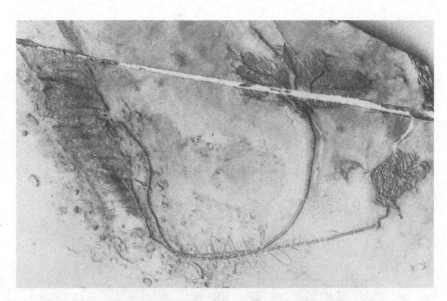

Figure 165 *Imitatocrinus gracilior* (F. Roemer, 1863) and *Bathericrinus semipinnulatus* (W.E. Schmidt, 1934) (to the right) attached to a shell of *'Orthoceras'* sp.. The slab is traversed by a tension crack infilled with quartz, Eschenbach–Bocksberg mine, Bundenbach (×0.4; HS 119).

Figure 166 A group of eight *Parisangulocrinus minax* W.E. Schmidt, 1934, attached to a stem-like structure, perhaps the sponge *Retifungus*, Eschenbach–Bocksberg mine, Bundenbach (×0.3; HS 130).

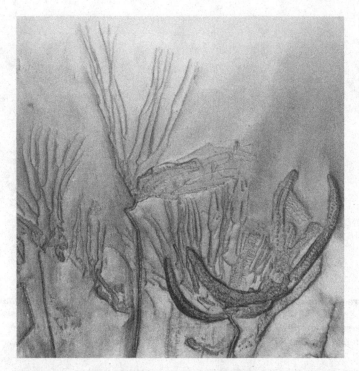

Figure 167 A group of six *Parisangulocrinus zeaeformis* (Follmann, 1887) in which an example of the asteroid *Urasterella asperula* has become trapped, Bundenbach (× 0.5; private collection).

Cyathocrininids (infraclass Cyathocrinina)

The Cyathocrinina are a monophyletic group, with ancestors among the cladids.

Codiacrinus

Codiacrinus is relatively common in some beds in the vicinity of Bundenbach. The cup was large, consisting of three infrabasal plates (partly concealed by the insertion of the stem), and five large basals and radials. There was no anal plate. The arms, which divided three times, were inserted in a notch in the radial plates. They are usually enrolled distally, beyond the third bifurcation. They bore short thorn-shaped plates, which presumably represent the position of pinnules. The ratio of width to height in the calyx increased with growth (Fig. 170), giving adult individuals of *Codiacrinus* a much more robust appearance than juveniles.

Flexibles (infraclass Flexibilia)

Like the cyathocrininids, the Flexibilia appears to be a monophyletic group, with ancestors among the cladids. The proximal parts of the arms were incorporated into the calyx, but not fused, so that the calyx was flexible. There were always three infrabasals, one smaller than the other two. Pinnules were absent.

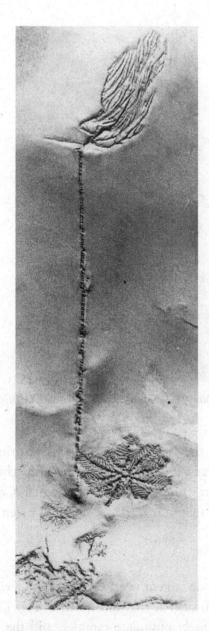

Figure 168 *Rhadinocrinus dactylus* W.E. Schmidt, 1934, with several juvenile specimens of *Thallocrinus* sp. near the base of the stem, Eschenbach–Bocksberg mine, Bundenbach (×0.6; HS 134).

Taxocrinus

Taxocrinus, one of the more common crinoids (Figs. 147, 171, 172), is the only flexible crinoid known from the Hunsrück Slate. The infrabasals were almost concealed by the insertion of the stem. The calyx was completed by five basals and five radials, together with the proximal plates of the arms. The arms divided

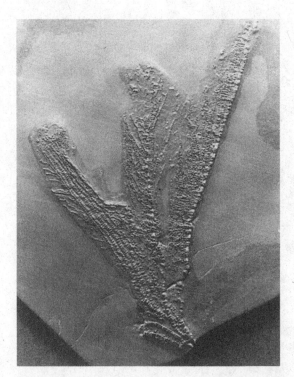

Figure 169 *Rhenocrinus ramosissimus* W.E. Schmidt, 1906, Bundenbach (×0.8; private collection).

four times. Distal of the third division they formed long thin terminations that often curve over the calyx. The stem was very long. Fragments approaching and sometimes exceeding 1 m in length are common at some horizons. Complete *Taxocrinus*, however, are very rare. The stems, at least, are usually broken during roof-slate production.

Undescribed crinoid

The crinoids of the Hunsrück Slate are in urgent need of systematic revision. A number of new forms await description, such as that illustrated in Fig. 173. The large calyx, with long arms bearing densely pinnulate ramules, and the characteristic variation in the size of the stem plates, are reminiscent of the camerates *Acanthocrinus* and *Diamenocrinus* (family Rhodocrinidae). There are significantly fewer free arms, however, and they are much shorter relative to the calyx.

Subphylum Eleutherozoa

Five major groups of Eleutherozoa are represented in the Hunsrück Slate, Edrioasteroidea, Asteroidea, Ophiuroidea, Echinoidea and Holothuroidea. Only the asteroids and ophiuroids occur in abundance. The others are rare.

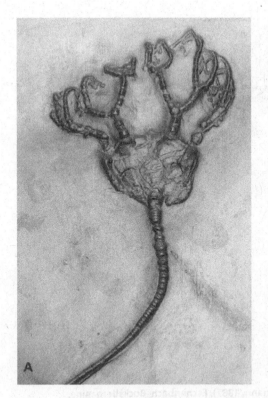 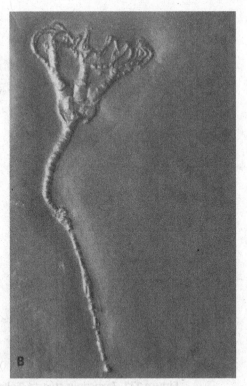

Figure 170 *Codiacrinus schultzei* Follmann, 1887: (A) Eschenbach–Bocksberg mine, Bundenbach (× 0.8; HS 125); (B) Bundenbach (× 1.0; SNG 091).

Edrioasteroids (class Edrioasteroidea)

The edrioasteroids were flattened or sac-shaped echinoderms with a partly flexible theca that was covered by numerous polygonal or rounded plates. The ambulacral grooves were usually curved and their appearance is reminiscent of the arms of an asteroid. The upper surface accommodated the mouth, near the centre, and the anus, in an interradial area where it was covered by a pyramid of small plates. A third opening, the hydropore, lay between these and was presumably part of the water vascular system.

Pyrgocystis

The only edrioasteroid known from the Hunsrück Slate is *Pyrgocystis (Rhenopyrgus) coronaeformis* (Fig. 174). The body was tower-like. The five ambulacra were restricted to the oral surface, which bore rows of spines. Most of the long body was covered with scale-like plates which overlapped and increased in size upwards. Proximally there was a sac-like 'foot', with numerous minute scattered plates, which presumably provided anchorage in the sediment (Rievers 1961*a*).

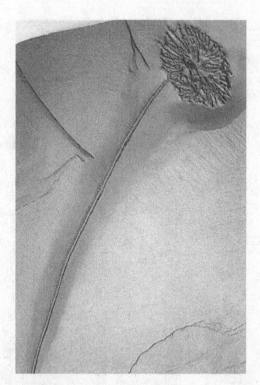

Figure 171 *Taxocrinus stuertzi* (Follmann, 1887), Eschenbach–Bocksberg mine, Bundenbach (× 0.3; HS 358).

Research on Hunsrück Slate edrioasteroids has been somewhat checkered. The original description of *Pyrgocystis* was based on a single well-preserved specimen discovered by Rievers. When he died before completing his research the manuscript was passed to Dehm who added further information before publishing it in 1961. Since then more than ten new specimens have been found. One was mentioned as a worm-like organism by Bartels and Brassel (1990, p. 88), who nevertheless noted the possibility that it might represent *Pyrgocystis* (1990, p. 123, 177). The same specimen formed the basis of a description of a so-called tubicolous animal by Fauchald and Yochelson (1990*a*). However, further inspection of both specimen and x-radiograph revealed the division of the body into a proximal sac, plated column, and disc-like upper surface. This surface displays remarkably preserved thecal structures which confirm its identity as *Pyrgocystis* (R. Haude, personal communication). More recently discovered specimens (see Fig. 174) are even better preserved and show additional details.

A second edrioasteroid was described by Dehm (1967) as *Hemicystites* (*Rieversidiscus*) *planus*. The specimen, however, simply consists of a small example of the asteroid *Urasterella* surrounded by a thin layer of calcium carbonate. Indeed,

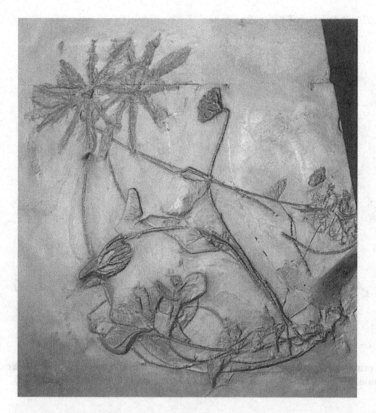

Figure 172 Group of two *Taxocrinus stuertzi* (Follmann, 1887) (with curled arms), two large *Imitatocrinus gracilior* (F. Roemer, 1863), one *Hapalocrinus* sp., and three undetermined juveniles, Eschenbach–Bocksberg mine, Bundenbach (×0.3; HS 530).

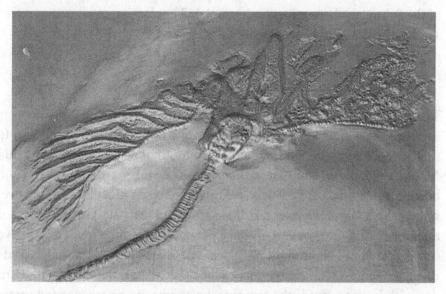

Figure 173 Undetermined rhodocrinid-like crinoid, Bundenbach (×0.9; SNG 089).

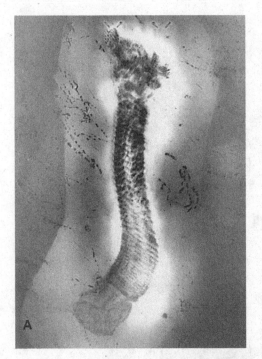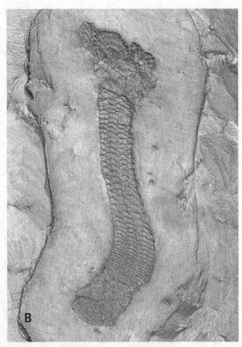

Figure 174 *Pyrgocystis (Rhenopyrgus) coronaeformis* Rievers, Eschenbach–Bocksberg mine, Bundenbach: (A) radiograph, (B) surface view (×0.9; HS 353, WS 12853).

Dehm noted (1967, p. 178) that the calcium carbonate alone would have been interpreted as an inorganic structure! Such white patches are common around fossils where the slates are influenced by meteoric water (in the Wingertshell layer of the Schmiedenberg mine, for example). The slate workers used to refer to them as 'Wasserplacken' ('water patches').

Asteroids (class Asteroidea)

The body of asteroids is flattened dorsoventrally. Isolated calcareous plates form a dermal skeleton. Most taxa have blister-like protuberances of the skin (papulae) that serve a respiratory function. The mouth is ventral, and the anus dorsal. Mouth-angle plates and first ambulacral plates form a frame around the mouth bearing spines that are used in feeding. The arms usually number five, but in some genera (e.g. *Palaeosolaster*) there are more than 20. On the lower surface of each arm is a double row of ambulacral plates, one tube foot lying between each successive pair. The ambulacral plates rest on adambulacral plates and beyond them, in some taxa, are marginal plates that define the edge of the arm. They diminish in size towards the arm tips. Lost arms can be regenerated provided that all of them have not been damaged. Most asteroids move by gripping the substrate with their tube feet.

Modern asteroids feed in a variety of ways. Some are detritus feeders or grazers, others are predators, feeding on gastropods, bivalves, small arthropods and even fishes. *Asterias* can pull the shells of bivalves apart by using the tube feet to grip them, and then evert the stomach to digest them. The functional morphology and mode of life of Paleozoic asteroids is poorly understood. In view of their apparent lack of suckered tube feet, and the typically large size of the mouth, Gale (1987) considered them to have been scavengers, deposit feeders, and predators on small prey. Recent research suggests that feeding in Paleozoic forms may have been similarly varied to that in modern forms. Mouth size is a function of geometry, and does not provide much indication of feeding habits. The presence or absence of suckered discs on the tube feet is uncertain, but some modern taxa are effective predators without these structures (Blake and Guensburg 1988). Evidence of extraoral feeding on bivalves has been described in the Ordovician asteroid *Promopalaeaster* (Blake and Guensburg 1994).

Modern analyses of echinoderm relationships (Smith 1984, Paul and Smith 1984) have shown that there is no basis for uniting the asteroids and ophiuroids in a group Asterozoa separate from other echinoderms. Hence they are treated as separate classes here. Very large individuals of several genera of asteroids and ophiuroids are known from the Hunsrück Slate, implying that conditions were particularly favourable for these animals.

Lehmann revised the asteroids of the Hunsrück Slate in 1957, describing 19 genera, 14 of which are now regarded as valid. Seven are illustrated here. Family assignments are those of Spencer and Wright (1966), which superseded Lehmann's treatment. The higher taxonomy of Paleozoic asteroids requires further investigation (Gale 1987, Blake 1994). They make up a stem group which lacks the advanced characters that unite the post-Paleozoic asteroids in the Neoasteroidea (Blake 1987, Gale 1987).

Baliactis

Baliactis (Figs. 175, 176) is one of the rarest of Hunsrück Slate fossils. Only two new specimens have been discovered since Lehmann (1957) described the Hunsrück Slate species. *Baliactis* is characterized by a broad plate (an interradial ossicle) that lies on the ventral side in each of the angles formed by the junction of the arms. The name *Baliactis* (from the Greek for tuberculate) refers to large shield-shaped plates on the dorsal side that show a tuberculate ornamentation.

A small central plate on the dorsal side of the body disc was surrounded by

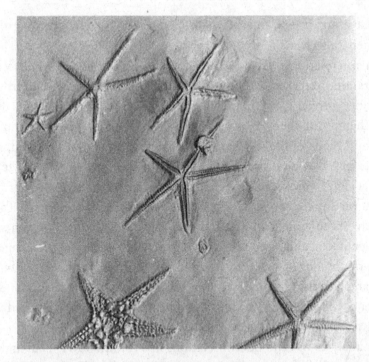

Figure 175 *Baliactis tuberatus* Lehmann, 1957 (the specimen on the left on the bottom margin) with six specimens of *Urasterella asperula* Roemer, 1863 (three with the dorsal and three the ventral side exposed), Eschenbach–Bocksberg mine, Bundenbach (× 0.7; HS 148).

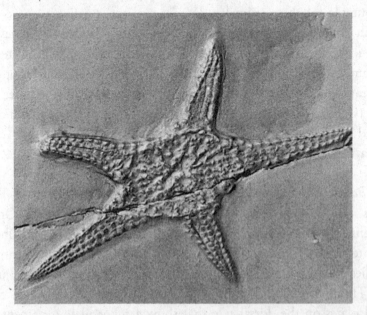

Figure 176 *Baliactis scutatus* Lehmann, 1957, Kaisergrube mine, Gemünden (× 0.8; SNG 134).

a ring of small plates covered with tubercles. Similar plates were irregularly distributed around this ring. The madreporite was subcircular, with a wrinkled surface. It lay between two arms near the margin of the body disc. A row of plates, each bearing a tubercle, ran along the mid-line of each arm continuous with a similar large plate near the border of the body disc. Spines, carried by elongate rectangular plates, formed a fringe around the distal part of the arm. On the ventral side a large shield-shaped interradial ossicle lay in contact with each pair of curved mouth-angle plates.

Fig. 175 shows *Baliactis tuberatus* associated with six individuals of *Urasterella asperula*. Lehmann (1957) described a second species, *B. scutatus*, on the basis of a single poorly preserved specimen. A new specimen from Gemünden (Fig. 176) is the only known complete example of this species, but it was only possible to expose the dorsal side. The dorsal plates are less developed than in *B. tuberatus*. *Baliactis* belongs to the family Taeniactinidae.

Echinasterella

Echinasterella (Fig. 177) was a medium-sized asteroid with five long slender arms. The dorsal skeleton consisted of a network of thorn-like plates which were more densely spaced toward the margin and bore tiny spines. The madreporite, which lay between two arms, was relatively large and slightly wrinkled. The mouth-angle plates each bore a long spine. The ambulacral plates were aligned and had a boot-shaped ridge on the surface. They lay in contact with the adambulacral plates which were similar and bore a lateral spine. There was a single Hunsrück Slate species, *E. sladeni*. *Echinasterella* is one of three Hunsrück Slate genera that belong to the family Helianthasteridae, which is characterized by a reticulate or granular dorsal surface and the presence of spines along the edge of the arms.

Helianthaster

This spectacular asteroid (Fig. 178) had 16 flat, ribbon-like arms. Some 15 to 20 specimens have been found in the last 20 years including some of the largest asteroids yet discovered, individuals with arms more than 20 cm long, indicating a total diameter of over 50 cm! The skin of *Helianthaster* contained calcareous granules. The central region bore some short spines. The areas of the body disc between the attachment of the arms had large scale-like plates that became thicker towards the margin. The madreporite was massive and wrinkled. The mouth frame, which is very obvious in the radiograph (Fig. 178), consisted of broad robust mouth-angle plates together with some ambulacral plates. The mouth-angle plates bore long spines that projected toward the centre of the

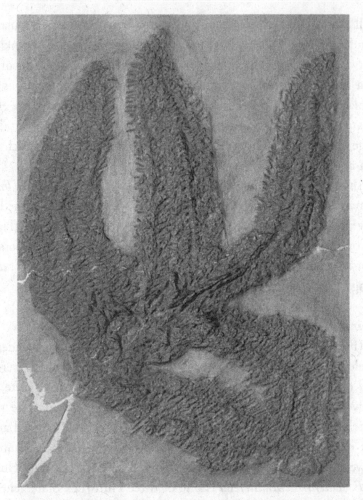

Figure 177 *Echinasterella sladeni* Stürtz, 1890, Eschenbach–Bocksberg mine, Bundenbach (× 0.9; HS 316).

body disc. The arm skeleton consisted of two rows of thin ambulacral plates flanked by slender adambulacral plates, each bearing a spine. Specimens of *H. rhenanus* are known with a smaller body disc and relatively short broad arms (the variant *microdiscus* of Lehmann 1957). *Helianthaster* belongs to the family Helianthasteridae.

Hystrigaster

This asteroid, which occurs only at Bundenbach, had a large body disc and five short broad arms with rounded tips. It is one of the large species in the Hunsrück Slate, normally ranging from 12 cm to 20 cm in diameter. The largest known individual is over 30 cm in diameter. The body was flat on the ventral side, and

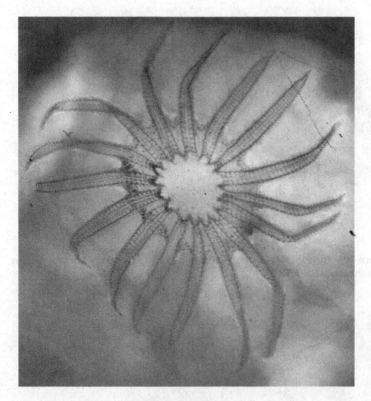

Figure 178 *Helianthaster rhenanus* Roemer, 1863, radiograph, Eschenbach–Bocksberg mine, Bundenbach (× 0.6; HS 576, WB 92).

presumably highly convex dorsally. The illustrated specimen (Fig. 179) is laterally compacted. The skeleton of the disc consisted of a network of elongate plates with small tubercles at the junctions between them. Five rows of long, robust spines with an ornamented surface were borne on these tubercles, articulating in a ball and socket joint. Similar spines were inserted on the skin between the skeletal elements. They were more densely spaced adjacent to the body disc. The subcircular madreporite, its surface covered in fine ridges and grooves, lay between two of the arms. The mouth was unusually large. The adambulacral plates were smaller than the ambulacral and bore tubercles with robust spines similar to those on the body disc. Just one species, *Hystrigaster horridus*, is known. *Hystrigaster* belongs to the family Helianthasteridae.

Palaeosolaster

Palaeosolaster was a large asteroid with 25 to 29 arms, the tips of which were rounded. The dorsal surface of the large body disc was covered by spines on many irregularly arranged small plates. The illustrated specimen (Fig. 180) had begun

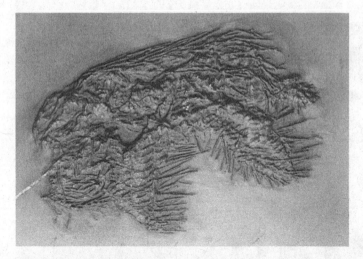

Figure 179 *Hystrigaster horridus* Lehmann, 1957, Eschenbach–Bocksberg mine, Bundenbach (× 0.6; HS 184).

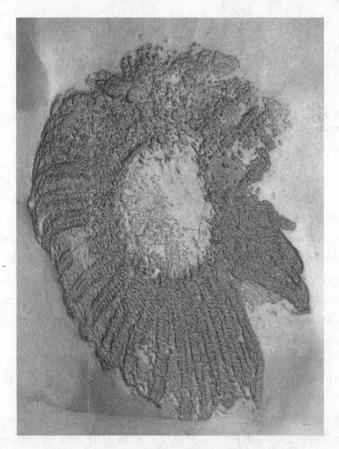

Figure 180 *Palaeosolaster gregoryi* Stürtz, 1899, slightly decayed and disarticulated, Eschenbach–Bocksberg mine, Bundenbach (× 0.4; HS 136).

to decay and some of the arms have disarticulated, revealing the large, elongate madreporite, which was covered by grooves and ridges. It lay on the dorsal side between two arms. The ambulacral grooves were wide and flat. The broad ambulacral plates had L-shaped ridges on the surface. The adambulacral plates, which were similar to the lateral plates of ophiuroids, formed the margin of the arms. They bore short spines. The large mouth was framed by narrow mouth-angle plates which show high relief. *P. gregoryi* is the only species of *Palaeosolaster* known from the Hunsrück Slate. *Palaeosolaster* belongs to the family Palasterinidae.

Palasteriscus

Palasteriscus is one of the rarest asteroids in the Hunsrück Slate. The illustrated example (Fig. 181) is one of the largest and most complete of about ten known specimens. Small rod-like plates covered the large body disc and formed transverse rows on the dorsal surface of the arms. The madreporite was very large and

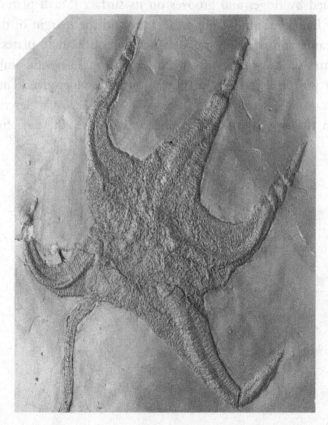

Figure 181 *Palasteriscus devonicus* Stürtz, 1886. Note the presence of *Chondrites* in the lower left. Eschenbach–Bocksberg mine, Bundenbach (× 0.5; HS 452).

lay between two arms on the oral side. The ambulacral plates were short and robust, and varied in position from aligned to alternating. The adambulacral plates were broad and low in relief, and bore short lateral spines. The mouth frame consisted of the mouth-angle plates and the first three to five ambulacral plates of each row. *P. devonicus* is the only known species of *Palasteriscus*. Spencer and Wright (1966) considered the family Palasteriscidae, to which *Palasteriscus* belongs, to be primitive asteroids, but Blake (1994) argued that the evidence for their place within the early phylogeny of the group remains equivocal.

Urasterella

Urasterella (Figs. 135, 175, 182, 183) accounts for 20–30% of all the specimens of asteroids and ophiuroids found in the Hunsrück Slate. Growth stages ranging in diameter from a few millimetres up to 30 cm are known. The body disc was very small. A hexagonal central plate was surrounded by five radial plates. These were surrounded by a further ring of small plates including the madreporite, which is distinguished by ridges and grooves on its surface. Small plates (paxillae) bearing tufts of hair-like spines were present around the margin of the body disc. The mouth frame consisted only of triangular mouth-angle plates. There were six rows of plates on the ventral surface of each arm. The wide ambulacral plates alternated in position, forming a narrow ambulacral groove. They bore slender spines on a transverse ridge. *U. asperula* (Fig. 182) was described by Roemer in 1863 in the first significant publication on the asteroids from the

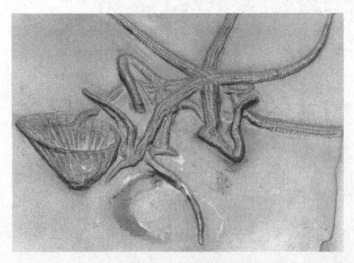

Figure 182 *Urasterella asperula* Roemer, 1863, three specimens with an unidentified rugose coral, Eschenbach–Bocksberg mine, Bundenbach (×0.8; HS 423).

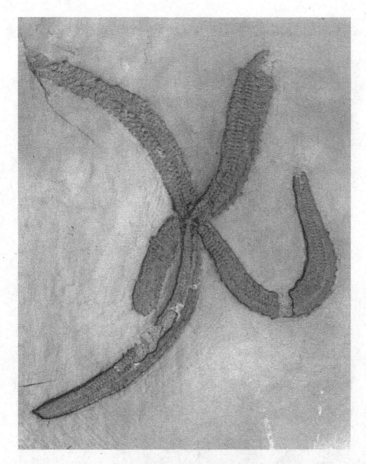

Figure 183 *Urasterella verruculosa* Lehmann, 1957. The tip of the arm at the top right has regenerated, Eschenbach–Bocksberg mine, Bundenbach (× 0.9; HS 504).

Hunsrück Slate. Lehmann (1957) described a rarer second Hunsrück Slate species, *U. verruculosa*, which is characterized by a granular skin on the dorsal side and a less robust skeleton (Fig. 183). *Urasterella* belongs to the family Urasterellidae.

Undescribed asteroid

The specimen illustrated in Fig. 184 appears to represent a new asteroid, discovered since Lehmann's work of 1957. This large compact asteroid is exposed from the oral side. The arms were broad, presumably highly convex, and covered by many irregularly arranged plates of variable size and shape. The specimen shows some similarity to *Echinasterella* but differs in lacking spines.

Juveniles

The presence of juveniles of many taxa in the Hunsrück Slate confirms that it represents a living environment and that oxygenated conditions normally pre-

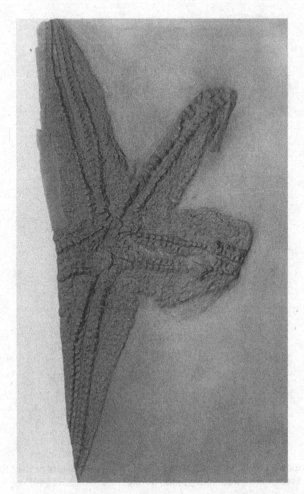

Figure 184 Asteroidea undetermined, Eschenbach–Bocksberg mine, Bundenbach (×0.8; HS 194).

vailed on the sea bottom. Fig. 185 is a much enlarged radiograph of a juvenile asteroid (the insert shows it about 1.2 times natural size). Examples less than 5 mm in diameter have been found. These fossils are usually too small to be determined taxonomically.

Holothurians (class Holothuroidea)

Holothurians (sea cucumbers) are elongate sac-shaped echinoderms. They are diverse in the modern oceans (there are about 1400 living species). Holothurians burrow or creep over the substrate using the suckered tube feet of the three ventral ambulacral areas. The other two ambulacral areas are dorsal in position with tube feet that function as sensory papillae. The mouth is surrounded by sticky

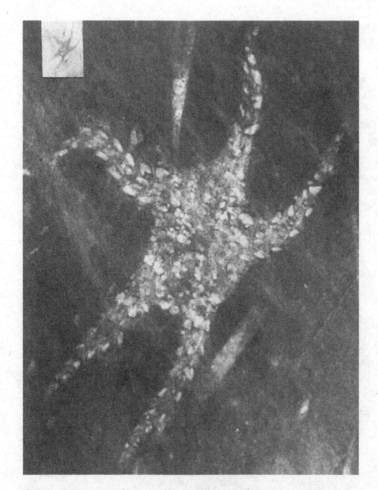

Figure 185 Juvenile asteroid, Bundenbach (×12.0, WS radiograph).

tube feet which are modified as tentacles for feeding. The anus is situated at the opposite end of the body. In contrast to other echinoderms the skeleton of holothurians consists only of isolated plates (sclerites) in the leathery skin. Hence fossil examples are rarely complete and usually consist only of dissociated sclerites (Gilliland 1993). Holothurians with a calcified skeleton first appear in the Lower Ordovician (Benton 1993), but the affinities of the Cambrian *Eldonia* may also lie with this group (see Briggs *et al.* 1994).

Palaeocucumaria

Palaeocucumaria (Figs. 186, 187), which is known only from the Hunsrück Slate, was discovered by Lehmann (1958*a*) as an image on an x-radiograph. The skin of *Palaeocucumaria* contained numerous small calcareous plates. They were more robust on the posterior part of the animal where they bore spine-like projections.

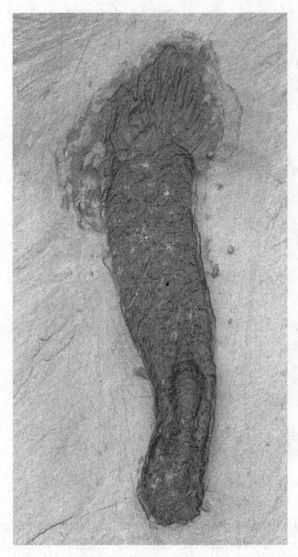

Figure 186 *Palaeocucumaria hunsrueckiana* Lehmann, 1958, Eschenbach–Bocksberg mine, Bundenbach (× 1.7; HS 533).

Anteriorly they decreased in size, becoming small grains near the base of the tentacles. *Palaeocucumaria* had about 20 tentacles which appear to have been arranged in pairs. Wrinkling gives them a segmented appearance and this was interpreted by Seilacher (1961) as evidence of internal calcareous elements (a feature that would be unique to *Palaeocucumaria* among holothurians: Frizzell and Exline 1966). There is, however, no evidence in x-radiographs (Fig. 187) for their presence. A ring of calcareous plates supporting the circle of tentacles is, however, evident. This ring, which Lehmann (1958*a*) mistook for a calcified ring

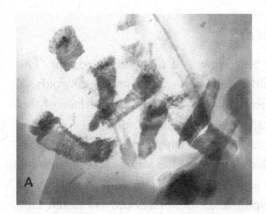

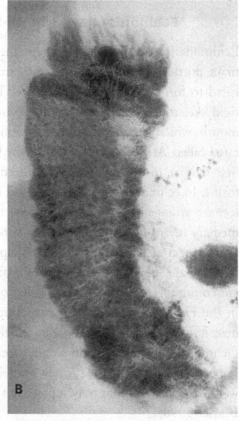

Figure 187 *Palaeocucumaria hunsrueckiana* Lehmann, 1958, Eschenbach–Bocksberg mine, Bundenbach: (A) group of individuals (× 0.5); (B) enlargement of the specimen on the left side (two isolated calcareous rings are evident on the right-hand side) (× 2.5) (Natural History Museum, Mainz PWL 1953/267-LS, WB 179).

canal, consisted of five robust radial plates with ten small plates lying between them (Seilacher 1961), an arrangement similar to that of living holothurians (Haude 1995). New specimens of *Palaeocucumaria* (Fig. 187) clearly reveal a large rounded structure in the centre of the crown of tentacles which is very reminiscent of the madreporite of many Hunsrück Slate asteroids (see Fig. 180). This represents the most recent example of a madreporite in holothurians; the characteristic element of the echinoderm water vascular system is absent in post-Devonian members of the class.

Complete specimens of *Palaeocucumaria* are rare (Seilacher 1961, Kutscher and Sieverts Doreck 1977) but they occasionally occur in clusters. Fig. 187A shows eleven individuals. Living holothurians can eject their inner organs in the event of trauma, and *Palaeocucumaria* may have reacted similarly when overwhelmed by a sudden influx of sediment. This may explain the isolated calcareous rings associated with the individual illustrated in Fig. 187B.

Echinoids (class Echinoidea)

Echinoids (sea urchins) are perhaps the best known of living echinoderms. The main portion of the skeleton (test), unlike that of asteroids and ophiuroids, is fused to form a spherical, flattened or heart-shaped 'corona'. The surface of this rigid skeleton is covered by many spines, large or small. In spherical forms the mouth, which is mid-ventral in position, is equipped with a formidable jaw apparatus called 'Aristotle's lantern' after the Greek philosopher who was reputedly the first scholar to observe it. This lantern is reduced or lost in different flattened forms. In echinoids, as in asteroids and ophiuroids, the plates and associated tube feet are arranged in five radiating ambulacral rows, which are separated by intermediary rows (interambulacrals). In spherical echinoids the anus is situated at the dorsal apex; in flattened and heart-shaped forms it lies in one of the interambulacral areas. Spherical echinoids are typically herbivores, whereas other echinoids are commonly detritus feeders. Paleozoic examples are rare, however, and those in the Hunsrück Slate are no exception. Three species are known. The original descriptions of *Porechinus porosus* Dehm, 1961*b* and *Rhenechinus hopstaetteri* Dehm, 1953 were based on single specimens from Bundenbach and Gemünden respectively (Kuhn 1961). A new individual of *Rhenechinus* is illustrated here, as well as a specimen that may represent a new species of *Lepidocentrus* (Bartels and Brassel 1990).

Rhenechinus

Unfortunately the edge of the new specimen of *Rhenechinus* (Fig. 188) was lost during the processing of the slate. The plates of this echinoid were imbricate and were not fused. There were four columns of plates in each ambulacral area and many regular columns in the interambulacral areas. The spines, which articulated on small tubercles, were more densely spaced on the ambulacra than the interambulacra. Fine ridges ran parallel to their long axis. The x-radiograph of this remarkably complete specimen clearly shows a large jaw or lantern within the test.

?Lepidocentrus

A less well preserved specimen (Fig. 189), belonging to the family Lepidocentridae, was discovered by Bartels in the Kreuzberg roof-slate mine of Weisel/Taunus. It is the first echinoid reported from the slates of that region. The test is somewhat distorted as a result of decay and disarticulation in the sediment. The side originally below (Fig. 189) remains essentially intact whereas some

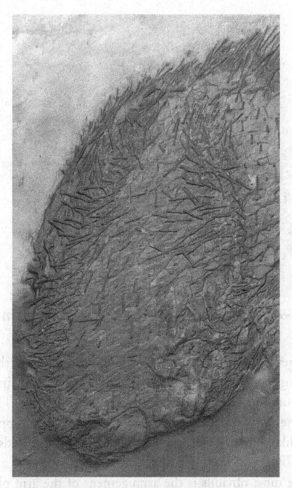

Figure 188 *Rhenechinus hopstaetteri* Dehm, 1953, Eschenbach–Bocksberg mine, Bundenbach (the right side of the specimen was lost during slate production) (× 1.0; HS 285).

scattering of the plates and spines has affected that above. The ambulacra, which appear to have consisted of two columns, are evident on the right and left side as illustrated (Fig. 189). Most of the spines, which were small and concentrated in the ambulacral areas, have separated from the test and are lying on its surface. The pentangular aperture surrounding the mouth is evident near the ventral margin. This echinoid may belong to the genus *Lepidocentrus*.

Ophiuroids (class Ophiuroidea)

Modern ophiuroids are characterized by long snake-like arms that emerge from a more or less circular body disc. Within the disc, the arms extend to the ventral

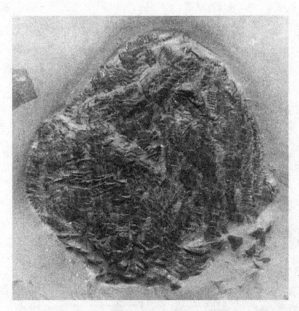

Figure 189 *Lepidocentrus*? sp., Kreuzberg mine, Weisel, Taunus (× 1.1; HS 337).

mouth. The arms are extremely flexible, their snake-like motion giving the ophiuroids their common name, 'serpent stars'. Many ancient ophiuroids, including typical representatives from the Hunsrück Slate, were very different. The discs were much larger, and the arms broader, and therefore their overall appearance was much more asteroid-like. Nevertheless, the Hunsrück Slate ophiuroids exhibit some characters that clearly reveal their affinities. Most of these are of a technical nature, but one of the most obvious is the arrangement of the arm plates. In ophiuroids the second row of plates from the axis of the arm (the 'laterals') lies lateral to the ambulacrals, whereas the second row in asteroids (the 'adambulacrals') lies at least partly under the ambulacrals. In living and most fossil ophiuroids the paired ambulacrals are fused to form a single plate, but this is never true of asteroids. The feeding habits of modern ophiuroids vary, but the large-disc forms from the Hunsrück Slate are so distinctive that their mode of life cannot be interpreted readily by analogy with the living forms.

Current classifications of ophiuroids are unsatisfactory and it is clear that a phylogenetic analysis is long overdue. The arrangement followed here is that adopted by Spencer and Wright (1966) and subsequently followed in Benton (1993). The faunal list includes 14 genera (23 species) of which 11 are considered here.

Cheiropteraster

Cheiropteraster (Fig. 190) was a rare and very large ophiuroid. Examples more than 40 cm in diameter are known. The skin that covered the body disc extended almost to the tip of the five arms. The calcite plates within this skin became larger toward the outer margin. The madreporite was small and round. The ambulacral plates were elongate and boot-shaped. The lateral plates were T-shaped and bore prominent flat spines along the edge. Tube feet were confined to the more proximal part of the arms. The mouth was very large and each of the mouth-angle plates bore a long slender spine which projected into the aperture. Only one species, *C. giganteus*, is known from the Hunsrück Slate. *Cheiropteraster* is one of four Hunsrück Slate genera that belong to the family Encrinasteridae, which is characterized by the presence of a frame of plates around the body disc and by the form of the lateral plates.

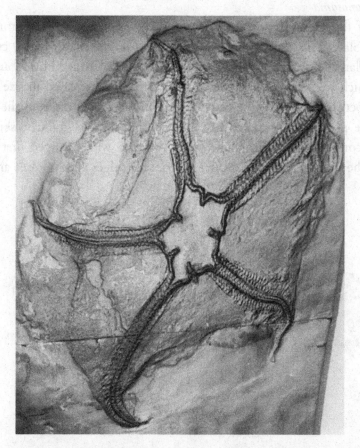

Figure 190 *Cheiropteraster giganteus* Stürtz, 1890, Eschenbach–Bocksberg mine, Bundenbach (× 0.3, private collection Bundenbach).

Encrinaster

This genus (Fig. 191) ranges extensively through the Paleozoic (from Ordovician to Carboniferous). The more common Hunsrück Slate species was originally described by Schöndorf (1910) as *Aspidosoma roemeri,* and subsequently assigned to *Encrinaster* by Schuchert (1914). The body disc was covered with plates which were larger toward the periphery and formed a thickened margin. The small madreporite was in the normal position. The arms were long and slender, and ended in flexible tips which extended beyond the large body disc. They were covered dorsally with a granular skin. The mouth frame consisted of robust mouth-angle plates together with enlarged first ambulacral plates. The ambulacral plates decreased in size distally, alternating in position. The lateral plates were rectangular and aligned at an oblique angle to the axis of the arm. *Encrinaster* belongs to the family Encrinasteridae.

Euzonosoma

One species of this genus, *Euzonosoma tischbeinianum* (Fig. 192), is common in the Hunsrück Slate. The largest specimens reached a diameter of over 40 cm. The body disc was large and covered by a granular skin. It was fringed by robust marginal plates which were smallest adjacent to the arms and increased in size toward the midpoint between them. These plates had pronounced tubercles in the adults. The arms were petal-like in appearance and covered with a granular skin. The mouth frame consisted of mouth-angle plates together with enlarged first ambulacral plates. The small madreporite was covered with ridges and grooves, and situ-

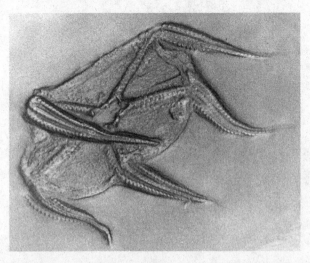

Figure 191 *Encrinaster roemeri* (Schöndorf, 1910), Eschenbach–Bocksberg mine, Bundenbach (× 1.0; private collection).

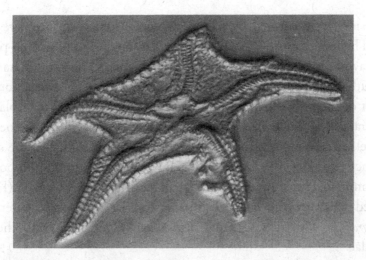

Figure 192 *Euzonosoma tischbeinianum* (Roemer, 1862), Eschenbach–Bocksberg mine, Bundenbach (×0.6, SNG 127).

ated in the normal position. The ambulacral plates were pentangular. The laterals were formed by the fusion of two components. *Euzonosoma* belongs to the family Encrinasteridae. This ophiuroid also occurs in slates of similar age near the Martelange Slate mine close to the Luxembourg–Belgium border.

Loriolaster
The body disc of *Loriolaster* (Fig.193) extended almost to the tip of the arms. It was very thin at the perimeter but became thicker toward the mouth. Both dorsal

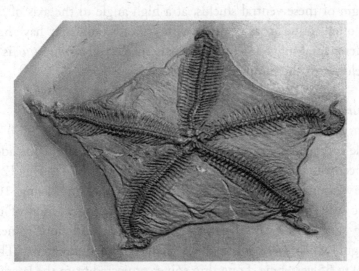

Figure 193 *Loriolaster mirabilis* Stürtz, 1886, Eschenbach–Bocksberg mine, Bundenbach (×0.7; HS 155).

and ventral sides were covered by a smooth skin. There were no plates on the dorsal side. The madreporite is not known even though dozens of well-preserved specimens have been studied; perhaps there was none. The mouth frame consisted of mouth-angle plates together with enlarged plates of the first ambulacrum. The skeleton of the arm consisted of broad boot-shaped ambulacral plates, and elongate lateral plates. These lateral plates bore small needle-shaped spines on the margin as well as one larger spine. These spines were embedded in the skin and probably functioned, together with the lateral plates, to support the body disc. The more common Hunsrück Slate species of *Loriolaster, L. mirabilis* (Fig. 193) was described by Stürtz in 1886; Lehmann (1957) described a rarer second species, *L. gracilis*, with more slender arms. *Loriolaster* belongs to the family Encrinasteridae.

Eospondylus

Eospondylus (Fig. 194), one of the more abundant ophiuroids, had five arms of medium length. The body disc was very small and incorporated only two or three of the arm segments. It was covered with small smooth imbricating scales. The madreporite was small and oval in outline. Opposing ambulacral plates were fused in pairs to form vertebrae which articulated in ball and socket joints. The individual ambulacral plates were boot-shaped. They were flanked by large sickle-shaped lateral plates. Ventral shields along the oral side of the arm could be moved together to close the ambulacral groove and protect the tube feet, as well as the delicate circulatory and nervous systems. A row of long spines projected from the outer margin of these ventral shields, at a high angle to the axis of the arm. *Eospondylus* is often found in association with crinoids, and may have had similar habitat preferences. Only one Hunsrück Slate species, *E. primigenius*, is known. *Eospondylus* belongs to the family Eospondylidae.

Kentrospondylus

Kentrospondylus decadactylus was first described by Lehmann (1957) on the basis of a single spectacularly preserved specimen (Fig. 195). Only one additional example has been discovered (Bartels and Brassel 1990, p. 142, Fig. 125). The genus was very similar to *Eospondylus*, differing mainly in having 10 arms. The body disc was small and rounded. It included seven to nine of the arm vertebrae. The skin of the dorsal surface was granular and bore spines about 5 mm in length which were inserted on small rounded tubercles. The arms consisted of 50 to 55 vertebrae. Long thin spines projected from the lateral plates. The extremity of the arms was rope-like. A long slender spine projected into

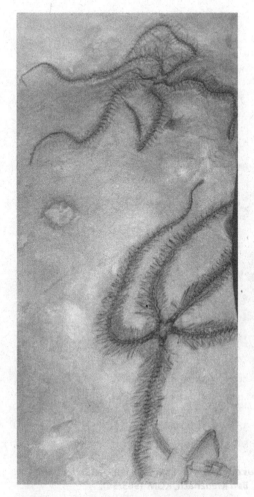

Figure 194 *Eospondylus primigenius* (Stürtz, 1886), Eschenbach–Bocksberg mine, Bundenbach, Haus-Plattenstein (× 0.6; HS 413).

the mouth aperture from each mouth-angle plate. *Kentrospondylus* belongs to the family Eospondylidae.

Furcaster

Furcaster is by far the most common ophiuroid in the Hunsrück Slate. It is rivalled in abundance only by the trilobite *Chotecops*. Hundreds of individuals occur on some bedding planes (one is displayed in the Natural History Museum in Mainz). The body disc of *Furcaster* was small. It was convex dorsally, and covered with a granular skin which bore small spines in places instead of granules. The very small, circular madreporite was situated near the mouth. The long slender arms terminated in a whip-like extremity. The ventral shields on the arms decreased

Figure 195 *Kentrospondylus decadactylus* Lehmann, 1957, Bundenbach, (×0.7, Schlosspark-Museum, Bad Kreuznach, KGM 1983/54).

Figure 196 *Furcaster palaeozoicus* Stürtz, 1886, Bundenbach (×0.5; SNG 132).

in size distally, alternating in position with the ambulacral vertebrae. They bore a row of needle-shaped spines. Each half of the ambulacral vertebrae was boot-shaped. They too bore clumps of spines. The mouth frame consisted of mouth-angle plates together with enlarged first ambulacral plates. Small spines projected from these plates into the aperture.

Fig. 22 shows a large group of *Furcaster palaeozoicus* that was buried by a rapid influx of sediment. The alignment of the specimens reflects the current that embedded them. Three or four arms of each individual are commonly out-stretched in the direction in which the current was flowing; one or two extend in the opposite direction. *F. palaeozoicus* (Figs. 136, 196) is the most common species. *F. decheni* (Figs. 197, 198) was larger and more robust than *F. palaeozoicus*; it reached a diameter of up to 0.5 m. The body disc was also larger relative to the length of the arms. *F. zitteli* (Fig. 199) differed from the other two

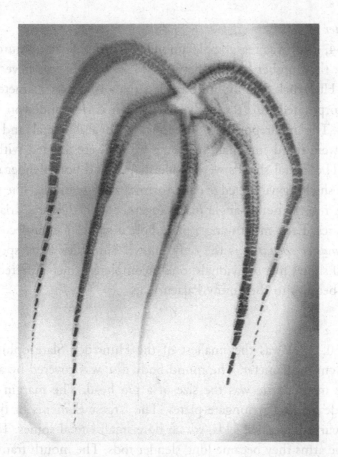

Figure 197 *Furcaster decheni* Stürtz, 1886. Late diagenetic euhedral pyrite has partly obscured the structure of the skeleton of the arms, Eschenbach–Bocksberg mine, Bundenbach (× 0.6, HS 575, WB 172).

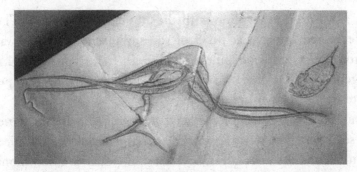

Figure 198 *Furcaster decheni* Stürtz, 1886, and *Weinbergina opitzi* R. and E. Richter (see Fig. 124), Eschenbach–Bocksberg mine, Bundenbach (× 0.3; HS 328)

Hunsrück Slate species in its robust mouth frame and broader arms. *Furcaster* belongs to the family Furcasteridae.

Medusaster

Medusaster (Figs. 34, 200) was remarkable among Hunsrück Slate ophiuroids in having many arms, the number varying between 11 and 16. It was nevertheless one of the smaller Hunsrück Slate ophiuroids, less than 10 cm in diameter. The arms were not incorporated into the body disc, which corresponded in size to the mouth frame. The madreporite is unknown. The ambulacral and interambulacral plates were fused into half vertebrae which were aligned with their opposite number. The lateral shields were sickle shaped and bore slender spines. The shape of these shields gave the edge of the arms a zigzag outline. The mouth frame included the robust mouth-angle plates together with the large triangular first ambulacral plates. Each mouth-angle plate bore a spine that projected into the aperture. *Medusaster rhenanus* is the only known Hunsrück Slate species of the genus. Fig. 200 shows five individuals, one incomplete, which suffered rapid burial. *Medusaster* belongs to the family Palaeuridae.

Ophiurina

Ophiurina (Figs. 50, 201) was the smallest of the Hunsrück Slate ophiuroids, rarely exceeding 5 cm in diameter. The round body disc was covered by a finely granular skin. The madreporite was the size of a pin head. The margin of the body disc was made up of 15 elongate plates. The largest elements of the arm skeleton were the curved lateral shields, which bore small lateral spines. Towards the extremity of the arms they became long slender rods. The mouth frame consisted of the mouth-angle plates and the enlarged first ambulacral plate. Fig. 201 shows two individuals associated with a specimen of *Taeniaster beneckei*.

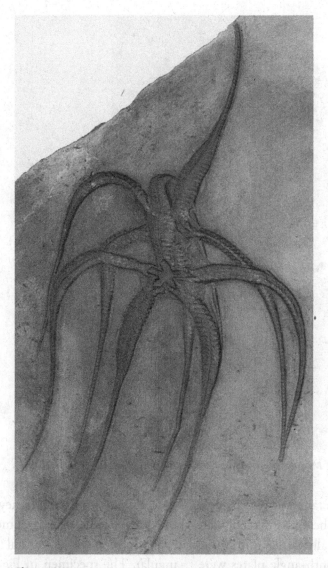

Figure 199 *Furcaster zitteli* Stürtz, 1886, Eschenbach–Bocksberg mine, Bundenbach (× 0.9; HS 390).

Ophiurina, like *Eospondylus*, is often found associated with crinoids, particularly *Bactrocrinites*. It belongs to the family Ophiurinidae.

Stuertzaster

The three Hunsrück Slate species now assigned to *Stuertzaster* are very rare. The body disc of *Stuertzaster* was medium sized and covered by granular skin. The arms were short and lancet shaped, flattened on the ventral side but convex dorsally. They were covered by a granular skin bearing many slender ornamented

Figure 200 *Medusaster rhenanus* Stürtz, 1890, Bundenbach (× 1.1; SNG 162).

spines. The ambulacral plates alternated with lateral plates to form a honey comb-shaped network. The outer lateral plates bore long needle-shaped spines. The mouth was unusually large and occupied nearly the entire ventral surface (Fig. 202). The mouth-angle plates were triangular. The specimen in Fig. 202 is probably the best preserved example of *Stuertzaster giganteus*. *S. tenuispinosus* (Fig. 203) differs in having large numbers of long spines on the arms. These species were placed in a new genus, *Erinaceaster*, by Lehmann (1957). Spencer and Wright (1966), however, demonstrated that *Erinaceaster* is synonymous with *Stuertzaster* Etheridge, 1899. *Stuertzaster* belongs to the family Pradesuridae.

Taeniaster
Taeniaster beneckei (Figs. 201, 204, 207) was one of the smaller ophiuroids, rarely exceeding 12 cm in diameter. It is common in the Hunsrück Slate. The body disc was oval and flat and covered by a coarsely granular skin on the dorsal side.

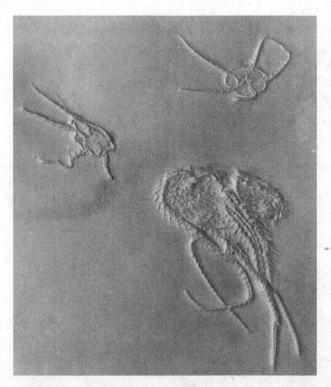

Figure 201 Two specimens of *Ophiurina lymani*, Stürtz, 1889, and one of *Taeniaster beneckei* (Stürtz, 1886), Eschenbach–Bocksberg mine, Bundenbach (× 1.0; SNG 164).

The madreporite, which was unusually small, was situated near the mouth frame. The arms were broad proximally, but narrowed into a long rope-like extremity. The arm skeleton consisted of alternating slender boot-shaped ambulacral plates which were flanked by robust adambulacral plates. These adambulacral plates bore delicate spines which were inserted between a ridge along the inner margin and a row of nodules on the outer margin. A larger, probably movable, spine also inserted on each adambulacral plate. *Taeniaster beneckei* was described by Stürtz in 1886 as belonging to a new genus *Bundenbachia*. Spencer (1934), however, demonstrated that *Bundenbachia* is synonymous with *Taeniaster*, which belongs to the family Protasteridae.

Undescribed ophiuroids

Since Lehmann (1957) described the Hunsrück Slate ophiuroids, several new forms have been discovered that have yet to be studied. Here we illustrate two of them with brief comments.

Fig. 205 shows a large ophiuroid similar to *Euzonosoma* with broad short arms that taper rapidly to a point. It differs from *Euzonosoma* in the way that

Figure 202 *Stuertzaster giganteus* Lehmann, 1957, radiograph, Eschenbach–Bocksberg mine, Bundenbach (× 0.7, private collection).

Figure 203 *Stuertzaster tenuispinosus* Lehmann, 1957, Eschenbach–Bocksberg mine, Bundenbach (× 0.7; HS 535).

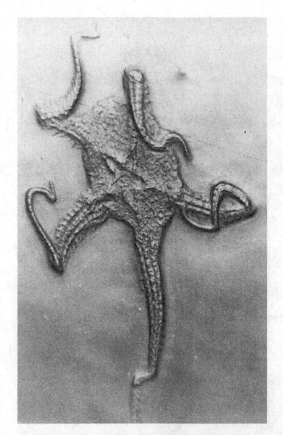

Figure 204 *Taeniaster beneckei* (Stürtz, 1886), Bundenbach (× 1.1; SNG 121).

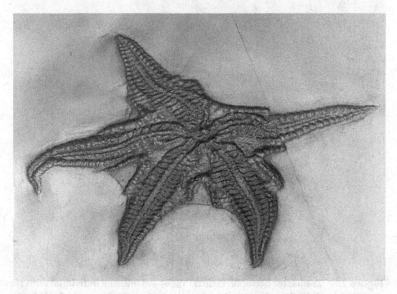

Figure 205 ?Euzonosomatidae nov., Eschenbach–Bocksberg mine, Bundenbach (× 0.7; HS 137).

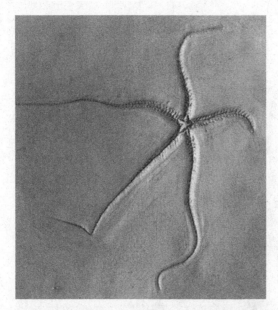

Figure 206 Ophiuroidea undetermined, Bundenbach (× 0.8; SNG 138).

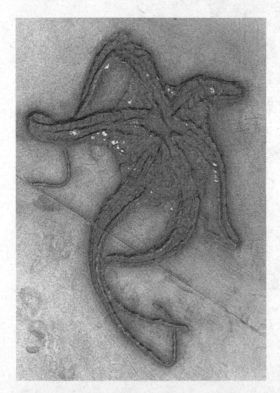

Figure 207 *Taeniaster beneckei* (Stürtz, 1886). An unusual individual with six arms. Late euhedral crystals of pyrite are evident on the surface, Eschenbach–Bocksberg mine, Bundenbach (× 1.1; HS 164).

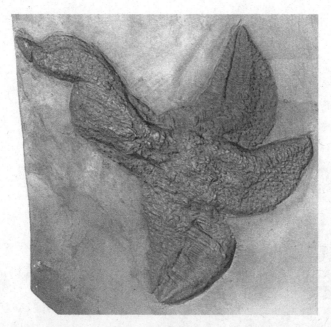

Figure 208 *Compsaster petaliformis* (Stürtz, 1886), an individual with only four arms, Eschenbach–Bocksberg mine, Bundenbach (× 0.6; HS 503).

the marginal plates do not follow the edge of the body disc but curve inward between the arms leaving an area of skin outside. Nevertheless this ophiuroid probably belongs to the same family Encrinasteridae.

Fig. 206 shows the oral surface of an ophiuroid. The body disc is very small and appears to include only the most proximal part of the long slender arms. Although this specimen resembles *Furcaster* and *Eospondylus* it is clearly different.

Abnormalities and injuries to the arms of ophiuroids and asteroids

Some of the Hunsrück Slate ophiuroids and asteroids had a variable complement of arms. The number in *Medusaster*, for example, varied between 11 and 16. Most genera, however, had five arms and only occasionally show a departure from this number. Examples of the ophiuroid *Taeniaster beneckei* are known with six arms (Fig. 207) and with only four (see Fig. 157). Among the asteroids abnormal individuals of *Urasterella asperula*, *U. verruculosa* and *Hystrigaster horridus* have been reported (Lehmann 1957, Bartels and Brassel 1990). Fig. 208 shows the only known specimen of *Compsaster petaliformis* with only four arms.

Ophiuroids are well known for their ability to cast off an entire arm or a fragment when attacked, and then regenerate the lost part. Several Hunsrück Slate specimens provide examples of this. Fig. 209 shows a specimen of *Furcaster* which

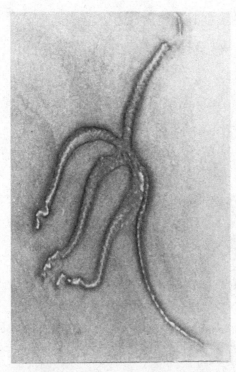

Figure 209 *Furcaster palaeozoicus* Stürtz, 1886. The tips of four of the arms of this individual have been injured, presumably by a predator, Kaisergrube mine, Gemünden (× 1.0; SNG 155).

suffered damage to the tips of four of the arms. It may have been attacked by some large arthropod such as *Palaeoisopus*. Damage to several arms would have reduced the animal's mobility by affecting the ambulacra and may have proved fatal. In asteroids it would also have resulted in the loss of the photoreceptors (optic cushions or eyespots) which are situated at the tips of the arms. The specimen of *Urasterella* illustrated in Fig. 183 has lost more than half of one of its arms and regenerated a short tip.

8 Vertebrates

The fossils of the Hunsrück Slate provide an unrivalled picture of the diversity of fishes in the early Devonian seas. The forms present include both agnathans and a range of gnathostomes, dominated by placoderms.

Agnathans

The agnathans are a paraphyletic group consisting of a range of extinct forms (that includes the ostracoderms) as well as the modern hagfishes and lampreys. These, the most primitive vertebrates, lack jaws. The ostracoderms had a dermal skeleton formed by bony plates that protected the head and trunk, while the tail was covered with scales. In the early forms, heterostracans like *Drepanaspis*, there was just a single gill opening on on each side of the body.

Drepanaspis

Drepanaspis gemuendensis is the only common agnathan in the Hunsrück Slate. It was the subject of a major study by Traquair (1903) and more recently by Gross (1963*a*). Over 100 specimens are now known (Kutscher 1973*a*) from Gemünden (the only source of articulated examples) and various other localities in the Middle Hunsrück, the Rhine Valley and the Taunus. A second agnathan, *Pteraspis*, is known only from two fragmentary examples.

The head and anterior part of the trunk of *Drepanaspis* (Fig. 210) was flattened. The large rostral plate just behind the mouth lacked any projecting rostrum. The orbital plates which surrounded the eyes were positioned anterolaterally and followed posteriorly by large postorbital plates. Much of the dorsal surface was covered by a single large plate with a characteristic outline (Fig. 211). An elongate branchial plate covered the gills. The posterior openings were flanked dorsally by

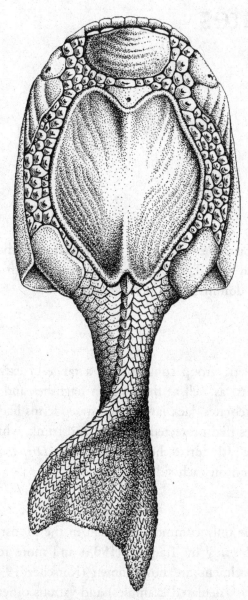

Figure 210 *Drepanaspis gemuendensis* Schlüter, 1887, restoration (after Moy-Thomas and Miles 1971).

a cornual plate. The surface of the larger plates was granular. Between these larger plates the head was covered by smaller elements (tesserae) with typical tubercle-like central thickenings. The trunk and tail were covered with robust scales, with a row of enlarged ridge scales along the dorsal axis. *Drepanaspis* ranges from 9.5 cm to 68.5 cm in length, but most of the specimens are between 35 cm and 45 cm.

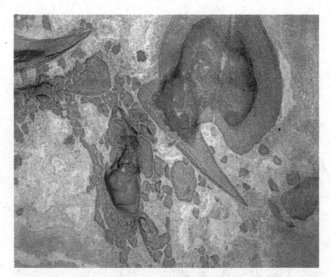

Figure 211 *Drepanaspis gemuendensis* Schlüter, 1887, partially disarticulated specimen, Kaisergrube mine, Gemünden (× 0.3; HS 359).

The highly flattened morphology of *Drepanaspis* suggests that it lived on the sea bottom. It is likely that the gills were ventilated simply by the action of forward swimming ('ram' ventilation) using the tail (Maisey 1996). The absence of paired or dorsal fins may have resulted in limited manoeuvrability compared with more advanced fishes. *Drepanaspis* presumably fed by using the broad slightly upturned mouth to pick up organic particles from the surface of the mud.

Placoderms

The gnathostomes are a monophyletic group characterized by the presence of jaws. The bizarre primitive forms known as placoderms, which were diverse in the Devonian, had a covering of bony plates on both the head and part of the body. Although they had various specializations of the jaws they lacked true replaceable teeth. The placoderms are represented in the Hunsrück Slate by several genera, but usually only as rare fragmentary specimens.

The majority of placoderms are classified as arthrodires (order Arthrodira). The most completely known Hunsrück Slate genus is *Gemuendenaspis* (Fig. 212) based on a single specimen that preserves most of the head and thoracic armour (Miles 1962). Another small arthrodire, *Stuertzaspis*, is known only from a single specimen of the narrow head (Westoll and Miles 1963). The largest Hunsrück Slate arthrodire is *Tityosteus* (Gross 1960). An example from Bundenbach, on display in Karlsruhe Natural History Museum (Otto 1992), may have reached a

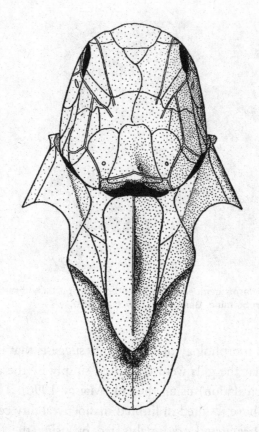

Figure 212 *Gemuendenaspis angusta* (Traquair, 1903), restoration (after Miles 1962).

length of 2 m. Fragments of large arthrodires similar to *Tityosteus* (Fig. 213) discovered in the Middle Hunsrück, south-eastern Eifel and Taunus areas (Bartels and Brassel 1990) await detailed investigation. Other placoderms from the Hunsrück Slate are more completely known, including *Lunaspis* (order Petalichthyida) and the ray-shaped *Gemuendina* (order Rhenanida).

Lunaspis

Lunaspis was named by Broili (1929*c*) who saw a resemblance to a crescent moon in the outline of the long curved lateral spinal plates. The genus is confined to Bundenbach and Gemünden, where several specimens have been found, some of them well preserved. The body of *Lunaspis* was flattened, with a large head covered by bony plates. These dermal bones display a characteristic ornament of concentric ridges which resemble the contours on a map. The large eyes, which were situated anterolaterally on the dorsal margin of the head, are particularly characteristic of this genus. The mouth was positioned ventrally at the anterior extremity. Large

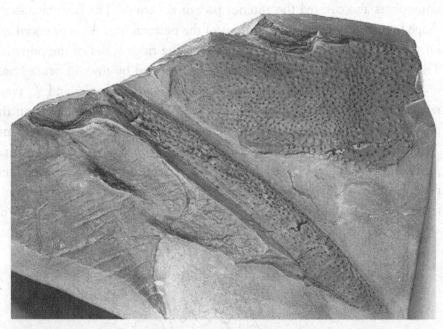

Figure 213 *Tityosteus rieversi* Gross, 1960, disarticulated skeletal elements, Mühlenberg mine, Bundenbach (× 0.6; HS 401).

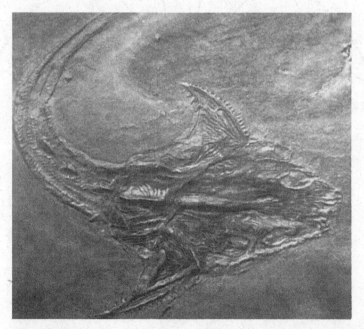

Figure 214 *Lunaspis heroldi* Broili, 1929 (× 0.8, Schlosspark-Museum, Bad Kreuznach KGM 1983/301).

bony plates also covered the anterior part of the trunk. The large characteristically-shaped spinal plates projected in front of the pectoral fins. A pronounced axial ridge on the median dorsal plate continued onto the ridge scales of the posterior trunk. The posterior trunk and tail were otherwise covered by rows of pentagonal scales.

The two Hunsrück Slate species of *Lunaspis*, *L. heroldi* and *L. broilii*, were revised by Gross (1961). They are distinguished by the ornament upon the spinal plates. Those of *L. broilii* are long, slender and sickle shaped, and bear more than 20 short spines on the anterior margin and six spines on the posterior margin. In *L. heroldi* (Fig. 214), in contrast, the spinal plates are shorter and more robust with no more than 15 spines along the anterior margin and four large spines on the posterior. *L. broilii* (Fig. 215) is the larger species, normally 30 cm to 45 cm long,

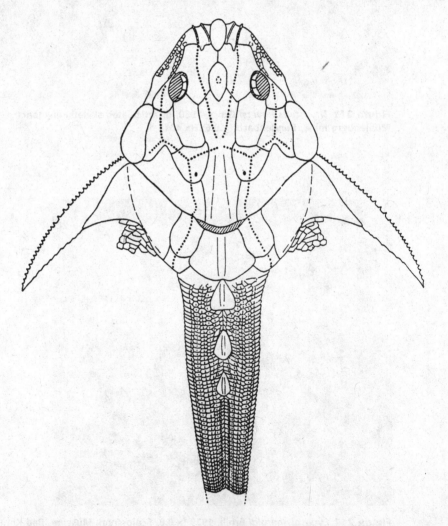

Figure 215 *Lunaspis broilii* Gross, 1937, restoration (from Gross 1961).

whereas some individuals of *L. heroldi* reach only 15 cm in length. The flattened body of *Lunaspis* was presumably an adaptation to life on or just above the sea floor.

Gemuendina

Gemuendina is the most completely known placoderm from the Hunsrück Slate. In appearance it is one of the most striking of the Hunsrück Slate fossils. The

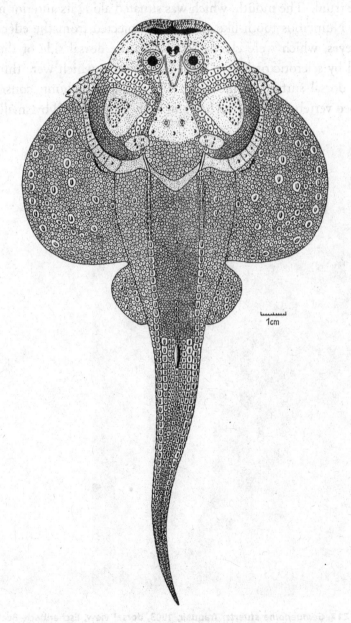

1cm

Figure 216 *Gemuendina stuertzi* Traquair, 1903, restoration (from Gross 1963*b*).

outline is reminiscent of a modern ray, particularly the large pectoral fins, the pelvic fins, and the long tapering flexible tail. The reconstruction by Gross (1963*b*) (Fig. 216) has contributed to its status as something of a Hunsrück Slate icon. All the known specimens are from Gemünden and Bundenbach, which has recently yielded some well-preserved individuals (Fig. 217).

The head of *Gemuendina* (Fig. 217) was roughly as long as wide, and large in relation to the trunk. The mouth, which was situated along its anterior margin, opened dorsally. Numerous tooth-like structures projected from the edge of the lower jaw. The eyes, which were closely spaced on the dorsal side of the head, were surrounded by sclerotic ossicles. Larger bony plates, which were thin, were confined to the dorsal surface of the head. The vertebral column consisted of simple ring-shaped vertebrae. Both the fins and body were covered by small scales.

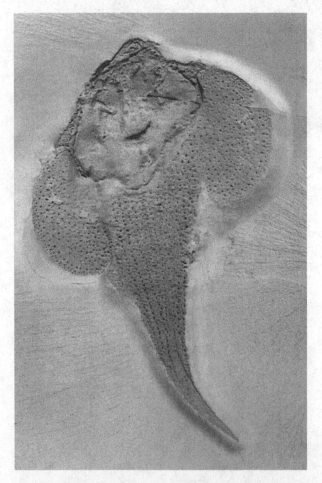

Figure 217 *Gemuendina stuertzi* Traquair, 1903, dorsal view, Eschenbach–Bocksberg mine, Bundenbach (× 0.3, private collection).

There was a dorsal spine in place of a fin. The majority of specimens of *Gemuendina* vary in length from 15 cm to 40 cm but one example nearly 1 m long is known from Bundenbach (Broili 1933*b*). Thus many of the individuals may be less than fully grown. The flattened form of *Gemuendina* suggests that it lived on the sea floor, swimming and feeding just above the substrate.

Other placoderms

The Hunsrück Slate has yielded four other placoderms: *Nessariostoma* (Fig. 218), *Stensioella* (Fig. 219), *Paraplesiobatis* and *Pseudopetalichthys*. *Stensioella* is known from just one nearly complete specimen (Gross 1962) and a head fragment; the other three genera from single specimens. These placoderms (Figs. 218, 219) were flattened like *Gemuendina*, but the body was narrower. The head was covered dorsally with dermal plates, and the body with scales. The delicate nature of the skeleton makes these fishes very difficult to detect in the slate prior to preparation. This may contribute to their rarity, and emphasizes the importance of careful treatment of possible discoveries.

Acanthodians

Acanthodians are a second major group of extinct gnathostomes, unrelated to the placoderms, that was very diverse during the early Devonian. They are very rare, however, in the classic localities of the Middle Hunsrück region.

Gross (1965) described material of *Machaeracanthus* (incorrectly listed as an arthrodire by Kutscher 1973*a* and Mittmeyer 1980*b*) from the Kaisergrube, and subsequently more acanthodian fragments have been found at this locality (Bartels and Brassel 1990). Acanthodians are unknown from the Hunsrück Slate of the Middle Rhine and Taunus regions, but disarticulated fragments (see Fig. 36) are widespread from the area around Mayen (Gross 1965, Bartels and Brassel 1990).

Figure 218 *Nessariostoma granulosa* Broili, 1933, Gemünden (Munich, Bavarian State Collection; (× 1.0 from Broili 1933*b*, plate 5).

Figure 219 *Stensioella heintzi* Broili, 1933, Eschenbach–Bocksberg mine, Bundenbach (× 0.7 from Broili 1933*b*, plate 3).

Here spines more than 40 cm long have been found. They vary in shape and size, some resembling curved horns, suggesting that a range of acanthodians is represented. Spines are also known from several localities in the Rhenish Normal Facies. All this material awaits detailed investigation.

The Eifel area yields large arthrodires in addition to acanthodians. This indicates that an abundant food supply was available for these large predatory fishes, perhaps including the large orthocones that are common in this region. Conditions in this part of the Hunsrück Slate, where asteroids and ophiuroids are absent, clearly differed from those in the classic Bundenbach area.

Sarcopterygerians (lobe-finned fishes)

Dipnorhynchus

A single specimen of a lungfish is known from the Hunsrück Slate. It was discovered near Bundenbach by Lehmann in 1945. The specimen is preserved in a very resistant matrix and its nature was only revealed by x-ray examination. This fossil, *Dipnorhynchus lehmanni*, represents the earliest record of a dipnoan fish (Lehmann and Westoll 1952, Lehmann 1956*c*).

Coprolites

The pyritized strings of faeces of deposit feeders are common at nearly all Hunsrück Slate localities. Coprolites (Fig. 220) that can be attributed to fishes,

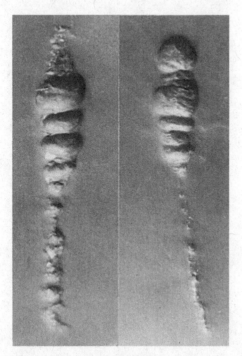

Figure 220 Fish coprolites, Moselberg mine, Irmenach, Hunsrück (left × 1.2; right × 1.0; private collection).

however, are abundant in only one area, around the villages of Hahn, Altlay and Irmenach, in the northern Hunsrück, where body fossils are very rare. This may reflect unusual conditions. The coprolites (Fig. 220) consist of solid crystalline pyrite and lack any organic remains that might yield information on the diet of their producers.

9 Trace fossils

Trace fossils preserve evidence of the activity of organisms living on and in sediment and therefore provide a useful indicator of environmental conditions. The preservation of soft tissues normally relies on the inhibition of scavengers and only occurs where the sediment is hostile to life. This has promoted the misconception that the preservation of trace fossils and that of soft-bodied fossils are mutually exclusive. Thus the evidence provided by trace fossils has tended to be neglected in studies of exceptionally preserved faunas (see Allison and Brett 1995). Trace fossils, however, are a ubiquitous feature of the Hunsrück Slate. Burrowing organisms were inhibited only in the immediate vicinity of the carcass. *Chondrites* may be associated with pyritized fossils (see Figs. 148, 181) although the burrows were presumably made after the onset of mineralization. Thus trace fossils from the adjacent sediment provide important evidence of the preservational setting.

Trace fossils have long been known from the Hunsrück Slate. The pioneering research was that of Rudolf Richter (1931, 1935, 1936, 1941) who based his interpretations of sedimentary structures and trace fossils from the Hunsrück Slate on comparisons with modern sediments, particularly the mudflats of the North Sea. It was Richter who first used the evidence of *Chondrites* to argue that the Hunsrück Slate sea was a living environment, not simply where animals died and became preserved. The trace fossils of the Hunsrück Slate were further discussed and illustrated by Seilacher and Hemleben (1966), who used them to argue a deeper depositional setting. A modern study of the ichnofauna by Owen Sutcliffe (Bristol) is underway with an emphasis on the taphonomy and environmental significance of the traces.

The diversity of trace fossils in the Hunsrück Slate is evidence of an established benthic community. A number of different types of behaviour are represented. These can be divided into the activities of organisms living on the sediment surface (epifauna) and those burrowing into it (infauna). The majority of the epifaunal traces were produced by arthropods walking or pushing off the substrate (see Figs. 221–224 below). Swimming fishes occasionally left traces on the surface of the sediment as a result of dragging their fins. Traces produced by walking ophiuroids are much rarer (Fig. 225). Sometimes locomotion was assisted by a current, and the trackways show alignment but little organization. Where the current is even stronger, arthropods and starfishes were transported passively and produced tool marks, sedimentary structures as opposed to trace fossils. Some epifaunal traces can be attributed to feeding or resting rather than locomotion.

Not all the Hunsrück Slate animals moved about on the surface of the substrate – locomotion and feeding within the sediment is also common (see Figs. 226, 227). Some animals moved from the surface into the sediment and vice versa (see Figs. 228–230). In many cases the trace maker cannot be identified, but a significant number of these trace fossils are the work of worms, bivalves and homolazoan echinoderms. The majority of infaunal traces, however, are complex open burrow systems made by unknown deposit-feeding organisms (see Fig. 231). These burrow systems have often been infilled from the surface with clay or silt. Many of them have a phosphatic lining, the result of early diagenesis associated with organic material, and are infilled with coarsely crystalline quartz and pyrite. The preponderance of open burrows suggests that the sediment in which the animals lived was dysaerobic or anaerobic so that contact with the seawater was essential. One of these burrow systems, *Chondrites*, is the most common trace fossil in the Hunsrück Slate (see Fig. 231). Evidence from a variety of lithologies of different ages elsewhere indicates that the animal that made *Chondrites* was tolerant of very low oxygen conditions (Bromley and Ekdale 1984).

The Hunsrück Slate animals were occasionally buried and killed by episodic influxes of sediment transported by turbidity currents. These turbidity currents also introduced fresh oxygenated sediment. Many of the traces record the activities of arthropods, worms and other animals sifting this sediment. Where oxygen levels remained elevated, the infaunal trace fossils are dominated by the simple burrow *Planolites*. Where oxygen has become depleted *Chondrites* is common.

Only a representative selection of the trace fossil taxa known from the Hunsrück Slate can be illustrated and described here.

Epifaunal traces

Arthropod trackways

Some of the most characteristic trace fossils from the Hunsrück Slate are those interpreted as the walking trails of the trilobite *Chotecops* (see Seilacher 1962). Each of the footprints is surrounded by impressions of fine setae or hairs (see Fig. 221). These setae fringed the distal extremity of the limb, to prevent the trilobite sinking into the sediment. Some other Hunsrück Slate arthropods, such as the crustacean *Nahecaris* (Bergström *et al.* 1987), had similar limb termina- tions and may have produced similar trackways. The amount of detail preserved in these and other trace fossils varies depending on the level at which the slate splits. Shallower imprints may be lost where the split runs below the surface on which the arthropod originally walked (so-called undertracking; Goldring and Seilacher 1971). Many of the arthropod trackways found in the Hunsrück Slate show much less organization than the walking trails of *Chotecops*, and clearly reveal the influence of currents. Fig. 222 shows where a trilobite impinged on the sediment surface, leaving an impression of the thoracic segments, before being swept away dragging its limbs behind. Bedding planes may be covered by the raking traces produced by the limbs of one (*Monomorphichnus*) or both sides of trilobites (*Dimorphichnus*), or even by the tips of the thoracic pleura, as the animals struggled to maintain contact with the surface of the sediment as they were dragged along by a current (Fig. 223).

A more complex arthropod trackway (Fig. 224) shows a series of imprints produced by brush-like limb tips pushing back into the surface of the sediment. It was first reported by Richter (1941, p. 238) on the basis of a single specimen. A second example, discovered by Bartels, was attributed to the spiny limbs of the large pycnogonid *Palaeoisopus* on the mistaken assumption that the broken slab

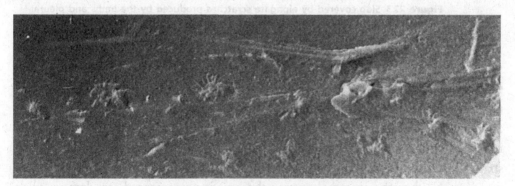

Figure 221 Trackway of *Chotecops*, showing the imprints of the fringe of spines around the extremity of the limbs, Kaisergrube mine, Gemünden (× 1.5; SNG 282).

Figure 222 The impression of a trilobite impacting the sediment surface and dragging its limbs as it was swept along by a current, Eschenbach–Bocksberg mine, Bundenbach (× 0.8; HSM 10).

Figure 223 Slab covered by elongate scratches produced by the limbs and pleural terminations of trilobites swept along by a current, Eschenbach–Bocksberg mine, Bundenbach (× 0.4; HSM 56).

preserved just one side of the trackway (Bartels and Brassel 1990, p. 100) whereas the whole is present. Specimens recently discovered at Bundenbach and Breitental, and in the collection in the University of Tübingen, show that this trackway is relatively abundant. The form illustrated (Fig. 224) grades into more widely spaced imprints as the arthropod takes off from the sediment surface and begins to swim. Neither the tracemaker, nor the activity represented, are known with certainty, but the arthropod may have been pursuing prey.

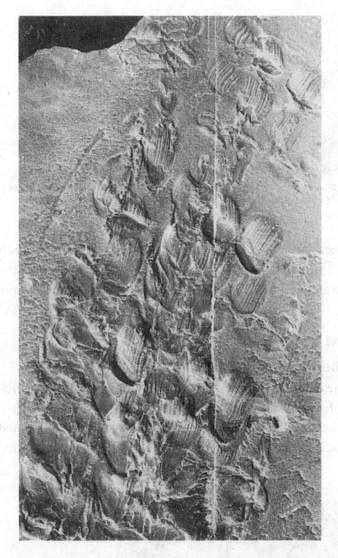

Figure 224 Trackway of an arthropod pushing the brush-like tips of its appendages into the sediment surface, Schmiedenberg mine, Bundenbach (× 1.0; HSM 5).

Ophiuroid trackway

Drag marks produced by ophiuroids carried passively in a current are known from the Hunsrück Slate (Seilacher 1959). Only one ophiuroid locomotion trail, however, has been reported (Bartels and Brassel 1990), a specimen recently described as *Arcichnus saltatus* (Sutcliffe 1997a). This trace fossil (Fig. 225) consists of successive horse-shoe shaped impressions that were produced by two arms, connected along the axis of the trackway by the imprint of a third arm. The trace is interpreted as the product of *Taeniaster* (see Figs. 201, 204) launching itself

Figure 225 *Arcichnus saltatus* Sutcliffe, 1997. The trackway of the ophiuroid *Taeniaster*, Eschenbach–Bocksberg mine, Bundenbach (× 0.5; HSM 1).

forward into a current with two anterior arms (Sutcliffe 1997*a*). It landed on a single posterior arm before falling forward to bring two anterior arms into contact with the substrate again, producing another horse-shoe shaped impression.

Infaunal traces

Simple burrows

Simple irregular meandering burrows are very common in the Hunsrück Slate. They vary from a few millimetres (Fig. 226) to more than 2 cm (Fig. 227) in diameter. The growth of early diagenetic pyrite or phosphate prevented compaction of these traces, so that they are preserved in relief. Pyritized examples are

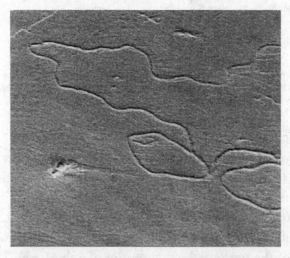

Figure 226 Meandering trace of a small worm, Eschenbach–Bocksberg mine, Bundenbach (× 1.4; HS/M).

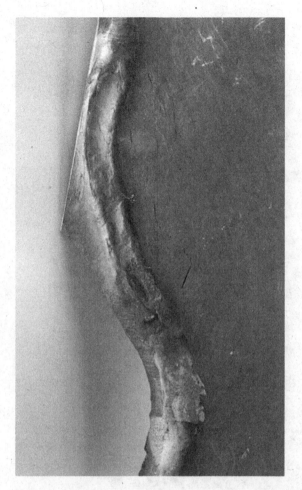

Figure 227 Large pyritized burrow, Eschenbach–Bocksberg mine, Bundenbach (× 0.6; HSM 59).

evident in x-radiographs where the small ones may be mistaken for crinoid stems. The trace makers are unknown, but were probably deposit-feeding worms.

Locomotion trails

A range of Hunsrück Slate trace fossils displays a series of paired imprints flanking a median area of disturbed sediment (Figs. 228–230). These traces were grouped together by Richter (1941) on the basis of their fancied resemblance to an ear of corn (Ährenförmige Fährte = ear-shaped trace). Some of them may have been initiated on the surface, but many show clear evidence of burrowing within the sediment and escaping through it following deposition (Figs. 228, 229). Their morphology varies not only with the identity of the trace-maker but also the nature of the sediment and degree of undertracking. Where the lateral imprints

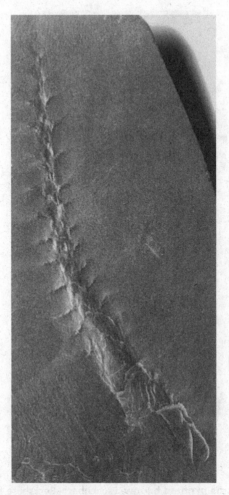

Figure 228 Trace produced by an arthropod with paired appendages moving through successive layers of sediment (Ährenförmige Fährte of Richter 1941 and Seilacher and Hemleben 1966), Eschenbach–Bocksberg mine, Bundenbach (× 1.0; SNG 279).

are clearly paired (Fig. 228), they resemble the trace fossil *Protovirgularia*, and may have been produced by the cleft foot of a protobranch bivalve (Seilacher and Seilacher 1994) moving through the sediment. Where this level of symmetry is lacking, however, other trace-makers may be implicated. More complex forms (Fig. 229) may have been produced by the paired limbs of an arthropod or the parapodia of a polychaete worm. The simplest examples (Fig. 230), which resemble arthropod walking trails like *Diplichnites*, may be the result of undertracking.

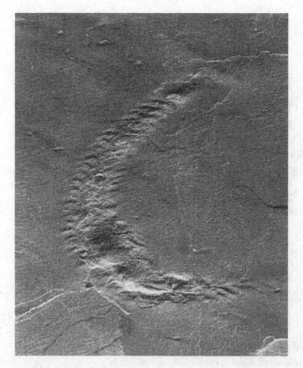

Figure 229 Similar trace to that illustrated in Fig. 228, but with more closely spaced limb impressions, Eschenbach–Bocksberg mine, Bundenbach (× 0.8; HSM 42).

Burrow systems

Chondrites

Chondrites is by far the most common Hunsrück Slate trace fossil. It is a branched burrow system (Fig. 231) with a distinctive root-like appearance. The diameter of the branches is uniform throughout, and they divide at angles of 30° to 40°. The burrow system is thought to have been kept open by the occupant, and subsequently filled with sediment from above (see Bromley and Ekdale 1984). Dense concentrations of *Chondrites* are present at many horizons (e.g. Fig. 16). A spectacular exposure of Hunsrück Slate displaying *Chondrites palaeozoicus* in the Leimbach Valley near Bacharach is protected as a Natural Monument (Kutscher and Horn 1963).

Heliochone

One of the most unusual burrow systems known from the Hunsrück Slate is *Heliochone hunsrueckiana* (Fig. 232). The structure of this trace fossil was unravelled by Seilacher and Hemleben (1966) by serially sectioning specimens. The burrow system normally consisted of a circular tunnel that was connected to the

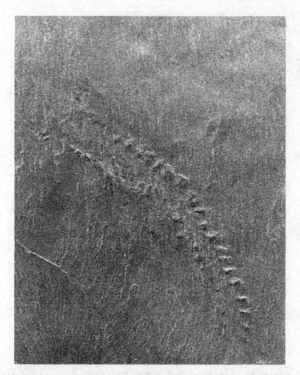

Figure 230 Trace showing the impressions of individual appendages that may differ from those in Figs. 228 and 229 as a result of undertracking, Eschenbach–Bocksberg mine, Bundenbach (× 1.3; HSM 11).

Figure 231 Restoration of *Chondrites* (from Simpson 1957).

surface at regular intervals by vertical shafts. As the animal exploited deeper levels in the sediment the diameter of the circular tunnel increased and consequently the vertical shafts migrated outwards. The maximum diameter of this trace fossil reaches 45 cm. The example illustrated (Fig. 232) is unusual in forming a spiral. The producer of *Heliochone* is unknown, but is thought to have been a deposit feeder (Seilacher and Hemleben 1966).

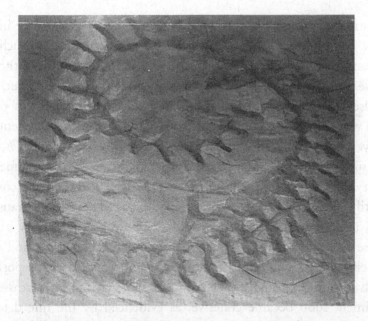

Figure 232 An unusual example of *Heliochone hunsrueckiana* Seilacher and Hemleben, 1966 showing a spiral form, Eschenbach–Bocksberg mine, Bundenbach (× 0.4; private collection).

Much larger structures, consisting of a series of regularly spaced vertical shafts projecting from a circular burrow, are evident on extensively exposed bedding planes in the Herrenberg Mine near Bundenbach, which is now a museum. These traces, which also occur in the Eschenbach–Bocksberg roof-slate mine, are 60 cm to 180 cm in diameter. They are simpler in structure than *Heliochone* and may represent complete examples of *Ctenopholeus kutscheri*, a trace fossil described by Seilacher and Hemleben (1966).

Environmental significance

The fine grained clay and silt that covered the Hunsrück Slate sea floor was ideal for the preservation of trace fossils. The abundance and diversity of epifaunal traces reflects the dominance of arthropods, particularly trilobites, among the body fossils. While the activities of these vagrant elements of the benthic community are well represented, the starfishes rarely leave traces. Their traces may not have been as clearly impressed or as distinctive. The makers of the infaunal traces are also incompletely represented among the body fossils. The rare pyritized polychaetes give an indication of the identity of larger burrowers, and other traces were produced by bivalves and homolazoans, but there is no evidence of the animals that made the complex burrow systems like *Chondrites* and *Heliochone*.

The location of the trace fossils within the sedimentary sequence indicates that a benthic community was established on the sea floor prior to each influx of sediment, and that the substrate was recolonized rapidly following the event. Some of these animals escaped through the newly deposited sediment, others came from elsewhere. This confirms the evidence of the body fossils that the water column was more or less continually oxygenated. Below the sediment–water interface, however, the abundance of complex burrow systems such as *Chondrites*, open to the surface, indicates that the normal condition of the sediment was anoxic, or at least dysaerobic (Bromley and Ekdale 1984). This condition was temporarily altered with each depositional event. The current-transported sediment was initially oxygenated and could therefore be invaded by a mobile infauna. These traces are a minority among the infauna, however, and do not penetrate much below the surface. Many may have been made by deposit feeders (e.g. bivalves, homolazoans) rather than animals that fed on carcasses. The iron-rich sediment soon became cohesive, as evidenced by the fine detail preserved in the surface traces, and anaerobic conditions were re-established, inhibiting any scavengers. Thus an unusual combination of circumstances on the Hunsrück Slate sea floor promoted the pyritization of soft-bodied fossils in association with an established benthic community.

Part III
Techniques and future research

10 Collecting and preparing Hunsrück Slate fossils

In the Rheinland–Pfalz region, which includes most of the outcrops of the Hunsrück Slate, fossils are classified as cultural or natural artefacts and enjoy the protection of the law. Excavations to collect fossils require the permission of the Office of Cultural Monuments (Landesdenkmalamt). Furthermore, the Landesdenkmalamt must be notified of discoveries of possible scientific importance because it is responsible for their protection and documentation. A law was passed in 1984 to control the activities of commercial collectors and prevent the destruction of unique sites in pursuit of rare fossils. The introduction of legislation was prompted by the use of heavy machinery to excavate in forest areas, for example, without permission. Nevertheless the vulnerability of even important fossil sites is demonstrated by the history of the famous Eocene Grube Messel near Darmstadt (Schaal and Ziegler 1988). It took 10 years of pressure from environmentalists, paleontologists and other interested parties to convince the government of Hessen that the protection of this world famous fossil site was more important than its potential as a refuse dump.

Collecting

It is very difficult to collect Hunsrück Slate fossils even by using heavy machinery to open up new excavations. Surface outcrops are weathered to a depth of several metres, resulting in the destruction of any specimens. The discovery of Hunsrück Slate fossils has therefore always depended on mining and quarrying for roof slate. The vast majority have come to light while the slate is being processed. The slate workers distinguish between 'Plattenstein', where bedding and cleavage are near parallel, and 'Krappstein', where the cleavage is at a significant angle to bedding. Plattenstein yields better fossil specimens and is softer and easier to prepare. The fossils in Krappstein, in contrast, often have associated pressure shadows, and are more distorted and even fragmented tectonically. Where quartz has been introduced its hardness makes preparation difficult. Sandier lithologies in the Hunsrück Slate are also more difficult to prepare, due to the recrystallization of the quartz grains.

The fossils of the Hunsrück Slate provide unique evidence of life in the past and it is essential that they are made available to specialists for study. Specimens of major significance must be deposited in scientific collections and properly documented. In this way amateur collectors make an important contribution both to scientific investigations and to the protection of our natural heritage. The acquisition of Hunsrück Slate fossils by museums and other scientific institutions has relied very largely on the work of amateurs. The role of protecting fossils from unscrupulous collectors must not be allowed to discourage the positive contribution to paleontology made by these private collectors. No less essential is the provision of adequate funding to support the on-going documentation of the geological setting of the remarkable Hunsrück Slate fossils. The Natural History Museum of Mainz is presently building up a collection from the Hunsrück Slate that promises to rival those elsewhere as a major repository of these fossils.

Collecting fossils in the Hunsrück Slate requires a different approach from that normally employed. Fossils may occasionally be found in the waste material resulting from roof-slate production, but this requires a good eye and a great deal of patience. It is also necessary to obtain the permission of the mine owners, and to observe the site regulations. Preparation of the surface of the slate is often necessary to establish whether or not a fossil is even present. Further time-consuming work is then required to expose the specimen. This need for preparation has been a problem for some public institutions that do not have the necessary equipment or personnel.

Preparation techniques

The Hunsrück Slate fossils do not separate easily from the matrix that encloses them. They rarely lie precisely parallel to the cleavage and are therefore never properly revealed by splitting the slate with a hammer and chisel. Wire-brushes made of steel or brass are used by some collectors as a quick method of uncovering fossils. Brushing destroys important surface detail, however, and damages delicate features such as the cirri of crinoids. Brass wire-brushes leave a golden residue (gilding) on the surface of the fossil, obscuring the natural colour of the pyrite. Although these 'golden' fossils are favoured by some collectors, the damage caused by this method of preparation renders the specimens largely useless scientifically.

Chemical methods have proved to be unsuitable. Hydrochloric acid removes the calcite that is an important component of most of the shelly fossils. Hydrofluoric acid dissolves the matrix but in so doing removes the support for delicate pyritized structures. Temperature change (heating and chilling, freezing and thawing) has proved ineffective in separating the matrix from the fossils, even though similar methods have been successfully used on other fine-grained sedimentary rocks (e.g. to expose the Carboniferous Mazon Creek fossils of Illinois). As Opitz (1932) remarked 'slabs [of Hunsrück Slate] burst open in all directions except the required one.'

There are only two reliable methods of preparing fossils from the Hunsrück Slate: (1) with hand-held needles and other scraping tools; and (2) with a specially constructed air-abrasive machine using fine-grained iron powder as abrasive. Only a few categories of Hunsrück Slate fossils are impossible to prepare. Very tiny fossils may be too fragile, in which case we have to rely on x-ray examination. Some soft tissues, such as the tentacles of cephalopods, are pyritized to such a limited extent that they are not coherent enough to resist mechanical treatment. The shells of molluscs, brachiopods, and some corals are usually not pyritized and any calcium carbonate that has survived dissolution during diagenesis is softer than the matrix. The matrix must be removed almost grain by grain to avoid damaging the shell, and the result is rarely satisfactory. Shells are much easier to prepare where the periostracum (in bivalves and brachiopods) has been pyritized. Trace fossils are usually impossible to prepare except where a contrast in lithology (and consequent hardness) allows the air-abrasive machine to be used.

Needle preparation

Preparation is very time consuming and requires a high level of concentration, accuracy and care. A good binocular microscope and appropriate illumination, as well as a comfortable working position, are essential (Fig. 233). The work demands a keen eye, a practised and delicate touch and, last but not least, a good ear! An experienced preparator of Hunsrück Slate fossils can hear quite clearly whether a tool is scraping over the slate matrix or the pyritized surface of the fossil! The preparator must be well acquainted with the anatomy of the organism to avoid removing delicate structures inadvertently. X-radiographs can be used to establish to what extent a specimen merits preparation and where the features to be revealed lie. The best preparation tools are usually modified for the purpose. Different types of needle can be filed and sharpened to provide a variety

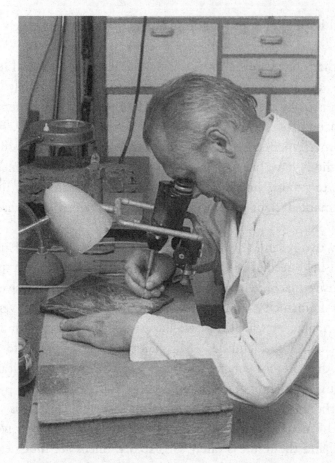

Figure 233 Günther Brassel using a needle to prepare a specimen of Hunsrück Slate crinoids.

of shapes and hardness, depending on their composition. Larger scrapers can be made from scalpels, small knives and files.

Fig. 234 shows the stages in the needle preparation of a specimen of *Loriolaster mirabilis*. Prior to preparation only the shadowy outline of the starfish is evident (Fig. 234A). The radiograph (Fig. 234B) clearly shows the structures to be revealed. First the matrix is scraped off until only a very thin layer remains overlying the specimen (Fig. 234C). Care must be taken to avoid removing any part of the pyritized fossil. At this stage the fossil is thoroughly washed and examined wet. Details are evident on a wet surface that are not visible when the slate is dry and may be too delicate to 'feel' with the scraper. Small scrapers and needles can now be used to expose details under the binocular microscope. Work begins in the mouth region and proceeds towards the tips of the arms (Fig. 234D). Once one arm is exposed, the preparator usually knows how to tackle the rest

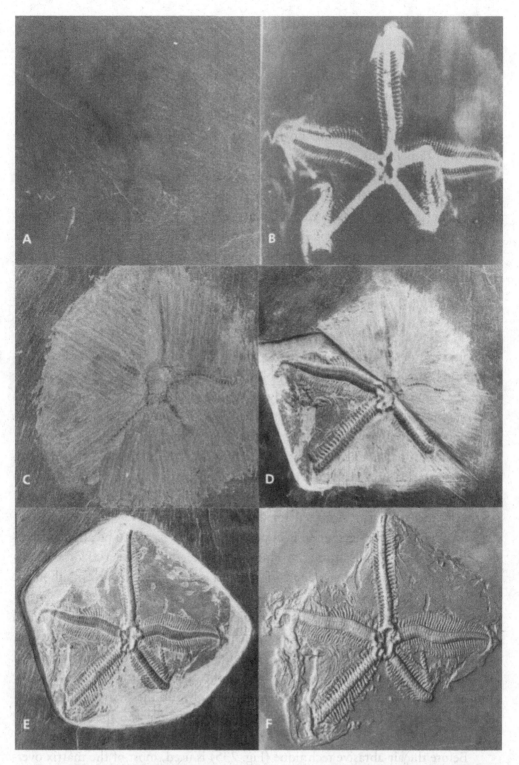

Figure 234 The preparation of a specimen of the Hunsrück Slate asteroid *Loriolaster mirabilis* by Gunther Brassel using needles and scrapers (× 0.7): (A) the unprepared slab showing the enclosed fossil in very faint relief; (B) x-radiograph supplied by W. Stürmer; (C) the specimen after some of the overlying matrix has been scraped away to show the outline of the mouth and arms; (D, E) stages in the preparation process; (F) the final result.

(Fig. 234E). When preparation is complete the fossil can be highlighted by smoothing the entire slab, or just a small area immediately surrounding the specimen (Fig. 234F). A drop of paraffin oil rubbed into the specimen seals and protects the pyrite from oxidation ('pyrite disease'). Finally the slab can be cut to the desired dimensions.

Fossils are often broken in the process of extracting and splitting the roof slate. The fragments must be glued together prior to preparation. Two-component glues or stone cements with a polyacryl composition are ideal (e.g. supplied by Akemi Products, Postbox 610163, Nürnberg). For delicate work, however, one-component acrylate glues (e.g. Superglue™) are preferable. Their low viscosity allows cracks to be repaired and fragmented slabs to be consolidated.

The air-abrasive method

Air-abrasive units, which are used to prepare fossils that are harder than the enclosing matrix, are like miniature sand-blasting machines. This method was first applied to Hunsrück Slate fossils nearly 30 years ago in Zürich by Kuhn-Schnyder (1969), who also experimented with the similar Eocene Glarus Fish Slate. Although the air-abrasive technique has since been adopted widely as a preparation method, it has not been extensively employed on Hunsrück Slate specimens. This is partly because air-abrasive machines are not part of the routine equipment of private collectors, but also because the method offered no improvement on the result or the time required for needle preparation.

Around 1990 fine elemental iron powder was discovered to be an ideal abrasive for preparing fossils from the Hunsrück Slate. H. Winkler (Hattersheim) constructed a new air-abrasive machine incorporating an innovative method for regulating the flow of powder to the specimen. Most air-abrasive machines introduce the powder into the airstream mechanically. It descends through fine-meshed vibrating sieves. However, these frequently clog up when flow is adjusted. Winkler constructed a tank for the powder that is connected to the source of compressed air. The rate at which powder is applied to the specimen can be regulated by adjusting the difference in pressure between the tank and the airstream that transports it. Thus the force and amount of powder reaching the specimen can be fine tuned to reflect the nature of the feature under preparation.

Before the air-abrasive technique (Fig. 235) is used, most of the matrix overlying a Hunsrück Slate fossil is scraped off (Fig. 235C), leaving just a thin skin. Jets 0.4–0.8 mm in diameter are best for preparing these fossils. Suitable pressures vary between 0.5 and 2.5 bars (higher pressures are unnecessary and can-

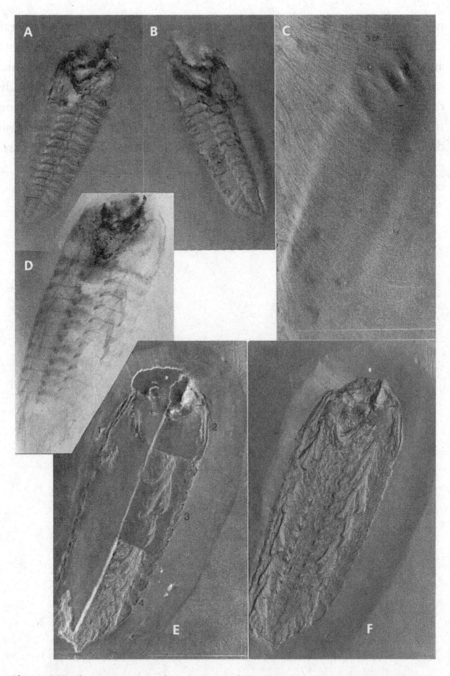

Figure 235 The preparation of a specimen of the Hunsrück Slate trilobite *Chotecops* sp. by Christoph Bartels using the air-abrasive method (A, B × 0.4; C–F × 0.6): (A, B) part and counterpart revealed as the slate was split; (C) the unprepared slab viewed from the ventral side after the part and counterpart have been glued together; (D) x-radiograph supplied by W. Blind; (E) stages in the preparation process – (1) the outline exposed; (2) the first features revealed; (3) after removal of more matrix appendages become visible; (4) part of the body completely exposed; (F) the final result.

not be used with jets of less than 1 mm). The air-abrasive tool is operated in a closed box with openings that allow the specimen and the jet to be manipulated, and the dust to be extracted. Viewing is facilitated by a binocular microscope installed on a swivel arm. The iron powder is heavier than the slate dust produced and over 99% of it can be recycled. The air-abrasive method reduces the time needed for detailed preparation to about 50% of that required using needles. Its major advantage, however, is that it allows delicate structures to be exposed that would otherwise be lost. Winkler's machine is not available commercially.

11 Modern analytical techniques and future research

X-radiography

Attempts to use x-rays to examine fossils followed very soon after their discovery by Wilhelm Conrad Röntgen in 1895. Hunsrück Slate fossils were used in the first experiments, which were made by Brühl in 1896 (Lehmann 1938, p. 16; Kutscher 1963a, p. 75). The first paper discussing the application of x-rays in paleontology was published in 1906 (Branco 1906). From the outset various shales and slates were the preferred subjects for these experiments (Branco 1906, Kutscher 1963a). More serious attempts to use x-rays to investigate the fossils from the Hunsrück Slate were initiated by Jaekel and Mautz in the 1920s, as reported by Lehmann (1938). Lehmann was the first paleontologist to apply x-ray techniques systematically to Hunsrück Slate fossils (Lehmann 1934, 1938, 1949, 1957, 1958a,b). He produced more than 90% of the 170 radiographs of Hunsrück Slate fossils published by the time of his death in 1959 (Kutscher 1963a). Stürmer (1970a,b, 1974, 1980, 1984, 1985) refined Lehmann's techniques to the standards of today. In so doing he discovered examples of spectacular preservation in the Hunsrück Slate including internal organs and previously unknown soft-bodied organisms (see Figs. 63, 65, 93). Stürmer systematically x-rayed slates in his specially equipped research vehicle in the search for fossils in the Kaisergrube at Gemünden. Several paleontological institutions now run modern radiographic laboratories.

Despite the spectacular images produced by x-radiography, it must be complemented by preparation of the specimens where necessary. X-rays may not reveal surface details that are of considerable importance taxonomically. Cracks infilled with quartz, which are clearly evident on radiographs (see Fig. 71B), cannot always be distinguished from the junctions between skeletal elements. Large euhedral pyrite crystals (see Fig. 197) may completely obscure important details on a radiograph that can nevertheless be revealed by preparation. Preparation of poorly pyritized fossils can reveal details that are invisible in the weak x-ray images that they produce. Specimens have been misinterpreted on the basis of x-radiographs alone (Fauchald and Yochelson 1990b). Last but not least, collectors and visitors to museums want to view the fossil itself, the tangible evidence of life millions of years ago, as well as the hidden structures revealed by x-rays (Fig. 236).

Since Stürmer's death in 1986 his x-ray equipment has been installed at the Institute of Applied Geology in Giessen, Germany, under the direction of Prof. W. Blind. The technique has been improved by using new high-resolution x-ray film (e.g. Agfa-Structurix™) which reduces the time necessary for an exposure from many hours (Stürmer 1984) to a few minutes (see Bartels and Blind 1995). Some of the new radiographs in this book (see Figs. 32, 37, 53, 88, 94, 100, 178, 187, 197, 234) were made under Blind's direction, demonstrating that much remains to be discovered in the Hunsrück Slate using this technique.

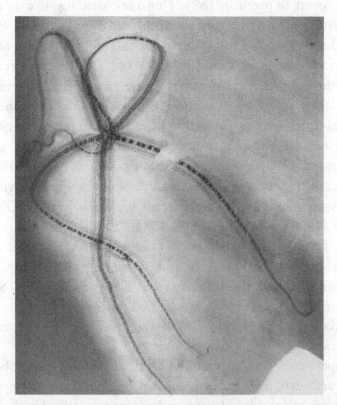

Figure 236 Radiograph of the asteroid *Protasteracanthium primus* showing variations in the degree of detail preserved depending on the nature of pyritization, Eschenbach–Bocksberg mine, Bundenbach (× 0.7, HS 577, WB 86).

Geochemistry

The first major investigation of the pyritization of the Hunsrück Slate fossils was undertaken recently to determine the factors controlling the remarkable preservation (see Briggs *et al.* 1996 for details of the techniques applied). The texture of the minerals was investigated with the scanning electron microscope (see Fig. 36). The nature of the mineralization process was interpreted by analysing not only the fossils, but also the sedimentary matrix that surrounds them. Samples of the slate were cut at intervals up to 12 cm laterally from a fossil, as well as above and below it. Both pyritized and non-pyritized specimens were analysed to allow a comparison of the concentrations of iron. All the samples were collected from newly quarried slate that was not affected by weathering. Iron was extracted using hot 12N HCl, and by dithionite, and analysed by Inductively Coupled Plasma – Atomic Emission Spectroscopy (ICP–AES). Concentrations typically increase towards the fossils, indicating that iron migrated into the decaying carcasses during diagenesis (Chapter 3). In addition iron concentrations are significantly higher in the slates that yield pyritized fossils, than in those that lack them. Analyses of iron can be used to calculate an expression for Degree of Pyritization (pyrite Fe/pyrite Fe + HCl soluble Fe) which provides an indication of how much oxygen was present in the bottom waters during deposition, and confirmed that the Hunsrück Slate sea was oxygenated.

Organic carbon was analysed in carbonate-free samples (obtained by treatment with 10% HCl) by measuring CO_2 by gas chromatography, following combustion, in an elemental analyzer. The carbon content of the slate has, of course, been modified by thermal and other diagenetic processes. Nevertheless the levels found in the Hunsrück Slate indicate that the original concentration in the sediment was low.

The relative timing of pyrite formation is revealed by sulphur isotopes. During the microbial sulphate reduction that leads to pyrite formation ^{32}S is more rapidly reduced than ^{34}S. Thus dissolved sulphide enriched in the lighter isotope is preferentially incorporated into the pyrite until the pore system becomes closed to the seawater by the overlying sediment. Following closure the proportion of ^{34}S progressively increases. A modified dental drill was used to extract pyrite from the fossils for sulphur isotope analysis. The isotopic signature of this fossil pyrite, and of samples of slate, was determined by extracting them with chromous chloride to produce CuS precipitates. Combustion of these precipitates with Cu_2O produced SO_2 gas that was then analysed on a mass spectrometer. The isotopic results show that precipitation of pyrite persisted for longer in association with the fossils than elsewhere.

Future research

The illustrated review of the Hunsrück Slate fossils presented in this book not only summarizes what is currently known about the biota, but also highlights where systematic revision of certain groups is needed and raises a number of fundamental questions that remain to be answered. The crinoids require a modern taxonomic treatment; they have not been studied since many were originally described in the 1930s by Schmidt (1934, 1941). The orthocone cephalopods are very poorly known. Smaller taxa, particularly microfossils, have been largely neglected. Trace fossils attracted some interest in the past (e.g. Richter 1935, 1936, 1941, 1954; Seilacher and Hemleben 1966) but only a fraction of the ichnofauna has been properly documented. The ichnology of the Hunsrück Slate is currently under investigation by Owen Sutcliffe in the Geology Department of the University of Bristol.

It is essential that contact between paleontologists and geologists and the slate-producing companies and their workers is maintained in the future. This will ensure that new discoveries continue to find their way to laboratories for scientific investigation. In addition there remains considerable potential for radiographic examination of the Hunsrück Slate fossils that are already in public and private collections; many important specimens have yet to be investigated in this way.

Research on the fossils of the Hunsrück Slate has been characterized by a lack of field investigation and systematic collecting at localities. This means that any wider conclusions are based on a very biased sample. The collections reflect the selection of the larger, more attractive and commercially valuable specimens by mineworkers. Thus any attempt to estimate the relative abundance of species based on existing collections is crude at best. No attempt has yet been made to analyse the distribution of all fossils at a single locality systematically. The impressions that we have gained during years of field collecting need to be tested in this way. This is a major objective of a new project, Project Nahecaris, organized by Michael Wuttke (Landesamt für Denkmalpflege Rheinland Pfalz, Mainz) together with Bartels and Briggs. This investigation will capitalize on a substantial thickness of fossiliferous roof slate which was extracted from the Eschenbach–Bocksberg quarry in 1997 with the aid of a 1.2 m diameter saw, specifically for scientific research.

The geology of the Bundenbach area, renowned for its fossils for over 100 years, has never been mapped! Mapping of the Gemünden sheet of the 1:25 000 Geological Map of Germany has yet to be initiated. There is only a simple sketch, published in 1984 in a popular description of the Hunsrück area (Kneidl 1984). Thus a fundamental tool for all geological investigations in this region is lacking. Only the roof slates of the Middle Hunsrück region around Bundenbach and Gemünden have been investigated extensively. The fossils themselves have been documented, but research on their preservational context has only been initiated recently with investigations of the stratigraphy, sedimentology and diagenesis (Briggs *et al.* 1996) at the Eschenbach–Bocksberg and Kaisergrube mines. Our knowledge of other localities in the Rhine valley, Taunus and Eifel is very incomplete. Much can be learned by continuing the investigation of old roof-slate mines and their dump material that was begun in the 1970s (Brassel and Kutscher 1978). Dozens of localities have never even been visited by a paleontologist. Thus many Hunsrück Slate outcrops are more or less *terra incognita*.

The first stage of the investigation of any major fossil deposit like that around Bundenbach is the description of new taxa and the interpretation of their environment and mode of life, data that contribute to our understanding of the evolution of life on earth. As more of the biota is documented, new organisms turn up less and less frequently. Prior to the 1960s, when Kutscher, Stürmer and others began to investigate the Hunsrück Slate, it was generally assumed that the potential of the deposit had been exhausted and that the slates were unlikely to provide significant new information. The 30 years since have shown this assumption to be wrong. Although new taxa are no longer discovered every time a paleontologist visits one of the roof-slate mines, research with modern methods, particularly x-radiography, has yielded many insights and, of course, raised many questions. Around 250 species of animal and 50 species of plant have so far been described from the typical roof-slate facies and the description and interpretation of new discoveries (Fig. 237) continues.

Future analyses of the mineralization of the Hunsrück Slate fossils will explore the relationship between the pyrite and other authigenic minerals. Evidence from the pyritized fossils of La Voulte-sur-Rhône in France (Wilby *et al.* 1996) indicates that the soft tissues were first mineralized in apatite before some were replaced by pyrite. A similar sequence may characterize the Hunsrück Slate fossils. The sequence of mineralization will be investigated by mapping the distribution of elements in the soft tissues on a very fine scale using an energy dispersive system (EDS) attached to a scanning electron microscope. It will also be instructive to analyse the sulphur isotope signature on a similarly fine scale using laser ablation techniques.

The pyritization of soft tissues will also be investigated in laboratory experiments. Pyrite has recently been precipitated in the laboratory in association with

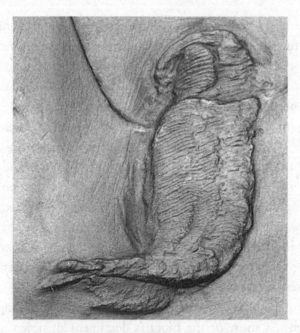

Figure 237 Worm-like segmented organism of unknown affinity, Eschenbach–Bocksberg mine, Bundenbach (× 1.4; HS 461).

the decaying carcasses of shrimps. The steep chemical gradients that develop in association with a decaying carcass can be monitored on a sub-millimetre scale using computer-controlled microelectrodes (Bale *et al.* 1996). This kind of approach offers the prospect of determining the exact role of iron, sulphur and oxygen concentrations, and pH, in pyritization. Thus the focus of future research on the Hunsrück Slate will be on the processes that led to the extraordinary preservation, and on how data from the biota can be used to address broader evolutionary questions.

Taxa recorded from the Hunsrück Slate

This taxonomic list is based on those of Mittmeyer (1980) and Bartels (1994), with additions and corrections. Only taxa found in the roof-slate facies are included.

Protists
Acritarchs gen. et sp. indet.

Plants
(Last systematic treatment Stürmer and Schaarschmidt 1980, excluding algae and spores)

Algae
Receptaculites cf. *neptuni*
(Defrance, 1827)

Tracheophyta
Prototaxites loganii Dawson

Spores (of primitive vascular cryptogams)
(Last systematic treatment Karathanasopoulus 1975)
Acanthotriletes simplex Naumova, 1953
Anapiculatisporites spinellosus
(Naumova) Lanninger, 1968
Anapiculatisporites sp.
Calamaspora cf. *atava*
(Naumova) McGregor, 1973
cf. *Chelinospora vermiculata*
Chaloner and Streel, 1968
Converrucosisporites sp.

Densosporites atavus De Jersey, 1966
Diaphanospora riciniata
Blame and Hassel, 1962
?*Dictyotriletes* cf. *eifeliensis* Schulz, 1968
Dictyotriletes minor Naumova, 1953
Emphanisporites annulatus
McGregor, 1961
Emphanisporites cf. *annulatus*
McGregor, 1961
Emphanisporites decoratus Allen, 1965
Emphanisporites cf. *epicautus*
Richardson and Lister, 1969
Emphanisporites erraticus
(Eisenack) McGregor, 1961
Emphanisporites protophanus
Richardson and Joannides, 1973
Emphanisporites rotatus McGregor, 1953
Emphanisporites schulzii McGregor, 1973
Emphanisporites sp.
Granulatisporites frustulentus
Blame and Hassel, 1962
Granulatisporites pipergranus
Haquebart and Barss, 1957
Granulatisporites sp.
Gravisporites asper Schulz, 1968
Leiotriletes atavus Naumova, 1953
Leiotriletes pagius Allen, 1965
Leiotriletes simplex Naumova, 1953

Leiotriletes cf. *sphaerotriangulus*
  (Loose) Pointone and Kremp, 1954
Leiotriletes sp.
Lophozonotriletes grandis Naumova, 1953
Lycospora cf. *pusillus* (Ibrahim) Schopf,
  Wilson and Bentall, 1944
Phyllothecotriletes golatensis Staplin, 1960
Punctatisporites cf. *curviradiatus*
  Staplin, 1960
Punctatisporites gemuendensis
  Karathanasopoulos, 1974
Punctatisporites punctatus (Ibrahim, 1933)
Reticulatisporites sp.
Retusotriletes communis Naumova, 1953
Retusotriletes lemniscatus Luber, 1955
Retusotriletes pychovii Naumova, 1953
Retusotriletes cf. *devonicus* Naumova, 1953
Retrusotriletes cf. *dubiosus*
  McGregor, 1973
Retusotriletes cf. *rotundus*
  (Streel) Lele and Streel, 1969
Retusotriletes cf. *subgibberosus*
  Naumova, 1953
Stenozonotriletes extensus var. *major*
  Naumova, 1953
Stenozonotriletes simplex Naumova, 1953
Synorisporites mikrogranulatus
  Karathanasopoulos, 1974
Tholisporites sp.
Verrucosisporites cf. *donarii*
  Pontonie and Kremp, 1954

Pteridophyta
Drepanophycus sp.
Hostimella sp.
Psilophyton sp.
Taeniocrada dubia
  Kräusel and Weiland, 1929
Taeniocrada sp.
cf. *Trimerophyton*

Animals

Phylum Porifera
(Systematics currently under revision by Kott,
Mehl and Wuttke)

?aff. *Choia* sp. (class Demospongia, order
  and family uncertain)
Asteriscosella nassovica
  Christ (order and family uncertain)
Dictyospongiidae gen. et sp. indet.
Retifungus rudens Rietschel, 1970 (order
  Reticulosa, family Dictyospongiidae)
Retifungus sp. (order Reticulosa, family
  Dictyospongiidae)
'*Protospongia*' *rhenana* (Schlüter, 1892)
  (order Reticulosa, family
  Protospongiidae)
'Rosselimorpha' fam., gen. and sp. indet.
  Mehl *et al.* 1997

Phylum Cnidaria
Class Hydrozoa
(Last systematic treatment Yochelsen *et al.*
1983)
  Plectodiscus discoideus
    (Rauff, 1939) (order Hydroida,
    suborder Chondrophorina, family
    Velellidae)
Class Anthozoa
(Last systematic treatment nineteenth
century)
  Aulopora sp. (order Auloporida, family
    Auloporidae)
  Pleurodictyum giganteum
    Kayser, 1880 (order Favositida,
    family Micheliniidae)
  Pleurodictyum hunsrueckianum
    Fuchs, 1899 (order Favositida, family
    Micheliniidae)
  Pleurodictyum problematicum
    Goldfuss, 1829 (order Favositida,
    family Micheliniidae)
  Pleurodictyum sp. indet. (order Favositida,
    family Micheliniidae)
  '*Rhipidophyllum*' *vulgare*
    Sandberger, 1847 (order Rugosa,
    family Zaphrentidae)
Rugosa gen. et sp. indet.
Zaphrentis primaeva Steininger, 1819
  (order Rugosa, family Zaphrentidae)
Class Scyphozoa
(Last systematic treatment Hergarten 1994)

Conularia bartelsi
Hergarten, 1994 (order Conulariida,
family Conulariidae)
Conularia bausbergensis
Hergarten, 1994 (order Conulariida,
family Conulariidae)
Conularia bundenbachia
R. and E. Richter, 1930 (order
Conulariida, family Conulariidae)
Conularia cf. *bundenbachia*
R. and E. Richter, 1930 (order
Conulariida, family Conulariidae)
Conularia gemuendina
R. and E. Richter, 1930 (order
Conulariida, family Conulariidae)
Conularia hunsrueckiana
Hergarten, 1994 (order Conulariida,
family Conulariidae)
Conularia mayensis
Hergarten, 1994 (order Conulariida,
family Conulariidae)
Conularia cf. *mediorhenana*
Fuchs 1915 (order Conulariida,
family Conulariidae)
Conularia tulipina
R. and E. Richter, 1939 (order
Conulariida, family Conulariidae)
Conularia cf. *tulipina*
R. and E. Richter, 1939 (order
Conulariida, family Conulariidae)
Sinusconularia blasii
Hergarten, 1994 (order Conulariida,
family Conulariidae)

Phylum Ctenophora
(Last systematic treatment Stanley and Stürmer
1983, Stanley *et al.* 1987)
Palaeoctenophora brasseli
Stanley and Stürmer, 1983
(family uncertain)
Archaeocydippida hunsrueckiana
Stanley, Stürmer and Yochelson,
1987 (family uncertain)

Phylum Mollusca
Class Gastropoda
(Last systematic treatment nineteenth century)

Loxonema obliquiarcuatum
Sandberger, 1889 (order
Mesogastropoda, superfamily
Loxonematacea)
Gastropoda gen. et sp. indet.
Class Bivalvia
(Last systematic treatment nineteenth century)
Buchiola bicarinata
Beushausen, 1895 (order
Praecardioida, family Praecardiidae)
Buchiola reliqua
Beushausen, 1895 (order
Praecardioida, family Praecardiidae)
Buchiola sp.
(order Praecardioida, family
Praecardiidae)
Ctenodonta aff. *insignis*
Beushausen, 1895 (order Nuculoida,
family Ctenodontidae)
Ctenodonta gemuendensis
Beushausen, 1895 (order Nuculoida,
family Ctenodontidae)
Cypricardella sp.
(order Veneroida, family
Cranadellidae)
Nuculites beushauseni
Fuchs, 1915 (order Nuculoida,
family Malletiidae)
Palaeoneilo bertkaui
(Beushausen, 1895) (order
Nuculoida, family Malletiidae)
Praecardium elegantissimum
Beushausen, 1895 (order
Praecardioida, family Praecardiidae)
Praecardium grebei
(Kayser, 1880) (order Praecardioida,
family Praecardiidae)
cf. *Rhenania tumida* Fuchs, 1915 (order
Trigonioida, family Myophoriidae)
Class Cephalopoda
Subclass uncertain
(Last systematic treatment Bandel and Stanley
1989)
Arthrophyllum
(= *Lamellorthoceras, Eoteuthis*)
Beyrich, 1850 (order uncertain,
family Lamellorthoceratidae)

Subclass Nautiloidea

(Last systematic treatment nineteenth century. Revision of the following *'Orthoceras'* and *Phragmoceras* species is overdue; some of them may belong to the family Lamellorthoceratidae)

> *'Orthoceras' digitale*
> > Sandberger, 1856 (order Nautiloidea, family uncertain)

> *'Orthoceras' percylindricum*
> > Sandberger, 1856 (order Nautiloidea, family uncertain)

> *'Orthoceras' planicanaliculatum*
> > Sandberger, 1856 (order Nautiloidea, family uncertain)

> *'Orthoceras' planiseptatum*
> > Sandberger, 1856 (order Nautiloidea, family uncertain)

> *'Orthoceras' sp.*
> > (order Nautiloidea, family uncertain)

> *'Orthoceras' tenuilineatum*
> > Sandberger, 1856 (order Nautiloidea, family uncertain)

> *Phragmoceras incertum*
> > Sandberger, 1856 (order Nautiloidea, family uncertain)

> *Phragmoceras subsulcatum*
> > Sandberger, 1856 (order Nautiloidea, family uncertain)

Subclass Ammonoidea

(Last systematic treatment Erben 1960–1965)

> *Anetoceras* aff. *hunsrueckianum*
> > Erben, 1964 (order Bactritida, family Mimosphinctidae)

> *Anetoceras arduennense*
> > Steininger, 1819 (order Bactritida, family Mimosphinctidae)

> *Anetoceras hunsrueckianum*
> > Erben, 1964 (order Bactritida, family Mimosphinctidae)

> *Anetoceras recticostatum*
> > Erben, 1964 (order Bactritida, family Mimosphinctidae)

> *Anetoceras (Erbenoceras)* sp. A, Erben, 1964
> > (order Bactritida, family Mimosphinctidae)

> *Anetoceras (Erbenoceras)* sp. B, Erben, 1964

> > (order Bactritida, family Mimosphinctidae)

> *Anetoceras (Erbenoceras)* sp. C, Erben, 1964
> > (order Bactritida, family Mimosphinctidae)

> *Anetoceras (Erbenoceras)* sp. D, Erben, 1964
> > (order Bactritida, family Mimosphinctidae)

> *'Bactroceras'* sp.

> *Cyrtobactrites?* sp.

> *Gyroceratites? laevis* (Eichenberg, 1931)
> > (order Bactritida, family Mimoceratidae)

> *Mimagoniatites falcistria* (Fuchs 1915)
> > (order Bactritida, family Agoniatidae)

> *Mimosphinctes* sp. (order Bactritida, family Mimosphinctidae)

> *Teicherticeras primigenium* Erben, 1965
> > (order Bactritida, family Mimosphinctidae)

Subclass Coleoidea

(Last systematic treatment Bandel *et al.* 1983)

> *Boletzkya longa*
> > Bandel, Reitner and Stürmer 1983 (order Boletzkyida, family Boletzkyidae)

> *Naefiteuthis breviphragmoconus*
> > Bandel, Reitner and Stürmer, 1983 (order Boletzkyida, family Naefiteuthididae)

> *Protoaulacoceras longirostris*
> > Bandel, Reitner and Stürmer, 1983 (order Aulacocerida, family Protoaulacoceratidae)

Class Hyolitha

Undetermined hyolithid, order and family uncertain

Class Tentaculitoidea

(Last systematic treatment Alberti 1982 *a, b*)

> *Nowakia barrandei*
> > Boucek and Prantl, 1959 (order Dacryoconarida, family Nowakiidae)

> *Nowakia hunsrueckiana*
> > Alberti, 1982 (order Dacryoconarida, family Nowakiidae)

> *Nowakia parapraecursor*
> > Alberti, 1982 (order Dacryoconarida,

family Nowakiidae)

Nowakia praecursor
Boucek, 1964 (order
Dacryoconarida, family Nowakiidae)

Nowakia sp. aff. *zlichovensis* Boucek, 1964
(order Dacryoconarida, family
Nowakiidae)

Styliolina hunsrueckiana
Fuchs, 1930 (nomen nudum?; order
Dacryoconarida, family Syliolinidae)

Styliolina sp. (order Dacryoconarida, family
Syliolinidae)

Viriatellina fuchsi
(Kutscher, 1931) (order
Dacryoconarida, family Nowakiidae)

Viriatellina gemuendina
(Runzheimer, 1932) (order
Dacryoconarida, family Nowakiidae)

Tentaculites grandis
Roemer, 1844 (order Tentaculitida,
family Tentaculitidae)

Tentaculites schlotheimi
Koken, 1889 (order Tentaculitida,
family Tentaculitidae)

Phylum Brachiopoda
(Last systematic treatment Mittmeyer
1973–1982)

Anoplotheca venusta
(Schnur, 1853) (order Spiriferida,
family Anoplothecidae)

Acrospirifer arduennensis antecedens
(Frank) (order Spiriferida, family
Delthyrididae)

Acrospirifer arduennensis prolatestriatus
Mittmeyer (order Spiriferida, family
Delthyrididae)

'Atrypa' lorana Fuchs
(order Spiriferida, family Atrypidae)

Brachyspirifer explanatus (Fuchs)
(order Spiriferida, family

Chonetes sarcinulatus (Schlotheim)
(order Strophomenida, family
Chonetidae)

Eodevonaria dilatata (F. Roemer)
(order Strophomenida, family
Eodevonariidae)

Euryspirifer assimilis s. str. (Fuchs)
(order Spiriferida, family
Delthyrididae)

Euryspirifer assimilis s.l. (Fuchs)
(order Spiriferida, family
Delthyrididae)

Hysteriolithes (Acrospitifer) arduennensis
Solle (order Spiriferida, family
Delthyrididae)

Plebejochonetes plebejus (Schnur)
(order Strophomenida, family
Delthyrididae)

Plebejochonetes semiradiatus (Sowerby)
(order Strophomenida, family
Delthyrididae)

Rhenorensselaeria demerathia
Simpson (order Terebratulida, family
Rhipidothyrididae

'Subcuspidella' incerta (Fuchs)
(order and family uncertain)

Phylum Bryozoa
(No systematic treatment except Brassel
1977)

Fenestella sp. (order Fenestrata, family
Fenestellidae)

Hederella sp. (order Cyclostomata, family
Hederellidae)

Phylum Annelida
(Currently under investigation)
Class Polychaeta

Bundenbachochaeta eschenbachensis
Bartels and Blind, 1995
(order and family uncertain)

Polychaeta gen. et sp. indet.

Sphenothallus sp. Fauchald *et al.*, 1986
(order and family uncertain)

Phylum Arthropoda
Subphylum Marrellomorpha
(Last systematic treatment Stürmer and
Bergström 1976)

Mimetaster hexagonalis (Gürich, 1931)
(order and family uncertain)

Vachonisia rogeri (Lehmann, 1955)
(order and family uncertain)

Subphylum Crustacea
Class Malacostraca
Subclass Phyllocarida
(Last systematic treatment Bergström *et al.*
1987, 1989)
>Ceratiocarina fam., gen. et sp. indet.
>*Dithyrocaris*? sp. (order Archaeostraca,
>>family Ceratiocarididae)
>*Heroldina rhenana*
>>(Broili, 1928) (order Archaeostraca,
>>family Ceratiocarididae)
>*Montecaris*? sp.
>>(order Archaeostraca, family
>>Rhinocarididae)
>*Nahecaris*? *balssi* Broili, 1930
>>(order Archaeostraca, family
>>Rhinocarididae)
>*Nahecaris stuertzi* (Jaekel, 1921)
>>(order Archaeostraca, family
>>Rhinocarididae)
>*Nahecaris* sp. (order Archaeostraca, family
>>Rhinocarididae)
>Rhinocarididae gen. et sp. indet.
>>(order Archaeostraca, family
>>Rhinocarididae)

Subphylum Arachnomorpha
Class Trilobita
(Last systematic treatment *Chotecops* Struve
1985, *Parahomalonotus* Brassel and Bergström
1978, others Richter 1931–1955)
>'*Asteropyge*' sp. (order Phacopida, family
>>Dalmanitidae)
>*Burmeisterella aculeata* (Koch, 1883) (order
>>Phacopida, family Homalonotidae)
>*Ceratocephala* sp.
>>(order Odontopleurida, family
>>Odontopleuridae)
>*Chotecops ferdinandi ferdinandi*
>>(Em. Kayser, 1880) (order
>>Phacopida, family Phacopidae)
>*Chotecops ferdinandi* (Em. Kayser 1880)
>>forma *propinqua* Struve
>>(order Phacopida, family Phacopidae)
>*Chotecops ferdinandi hypsipedops* Struve,
>>1985 (order Phacopida, family
>>Phacopidae)
>*Chotecops ferdinandi kutscheri* Struve, 1985

>>(order Phacopida, family Phacopidae)
>*Chotecops ferdinandi* (Em. Kayser, 1880)
>>forma *argus* Struve
>>(order Phacopida, family Phacopidae)
>*Chotecops opitzi* Struve, 1985
>>(order Phacopida, family Phacopidae)
>*Chotecops*? sp. *forma pseudargus*
>>Struve, 1985
>>(order Phacopida, family Phacopidae)
>*Cornuproetus hunsrueckianus*
>>E. Richter, 1936
>>(order Proetida, family Proetidae)
>*Dipleura* aff. *laevicauda* (Quenstedt)
>>(order Phacopida, family
>>Homalonotidae)
>'*Homalonotus*' *hunsrueckianus* Fuchs, 1899
>>(order Phacopida, family
>>Homalonotidae)
>'*Homalonotus*' sp.
>>(order Phacopida, family
>>Homalonotidae)
>*Odontochile rhenanus* (Kayser, 1880)
>>(order Phacopida, family
>>Homalonotidae)
>*Parahomalonotus planus* (Koch, 1883)
>>(order Phacopida, family
>>Homalonotidae)
>*Rhenops*? *limbatus* (Schlüter, 1881)
>>order Phacopida, family
>>Dalmanitidae)
>*Rhenops* cf. *anserinus* (order Phacopida,
>>family Dalmanitidae)
>*Scutellum wysogorskii* Lehmann, 1941
>>(order Ptychopariida, family
>>Thybanopeltidae)
>*Treveropyge drevermanni*
>>(R. Richter, 1909) (order Phacopida,
>>family Homalonotidae)

Superclass Cheliceriformes
Class Xiphosura
(Last systematic treatment Stürmer and
Bergström 1981)
>*Weinbergina opitzi*
>>R. and E. Richter, 1929
>>(order Xiphosurida, suborder
>>Synziphosurina, family
>>Weinberginidae)

Class Eurypterida
(Last systematic treatment Lehmann 1956)
  Rhenopterus diensti Størmer, 1939
      (order Eurypterida, family
      Rhenopteridae)
Class Scorpionida
  Palaeoscorpius devonicus Lehmann 1944
      (order Protoscorpiones, family
      Palaeoscorpiidae)
Class uncertain
(Last systematic treatment Stürmer and
Bergström 1978)
  Cheloniellon calmani Broili, 1932
      (order and family uncertain)
Class Pycnogonida
(Last systematic treatment Bergström *et al.*
1980)
  Palaeopantopus maucheri
      Broili, 1929
      (order Palaeopantopoda, family
      Palaeopantopodidae)
  Palaeoisopus problematicus
      Broili, 1928
      (order Palaeoisopoda, family
      Palaeoisopodidae)
  Palaeothea devonica
      Bergström and Stürmer, 1980 (order
      Pantopoda, family
      uncertain)

Phylum Echinodermata
Homalozoans
(Under revision by Jefferies and colleagues)
Class Stylophora
  Mitrocystites? styloideus
      Dehm, 1934 (order Mitrata, family
      Mitrocytitidae)
  Rhenocystis latipedunculata Dehm, 1932
      (order Mitrata, family
      Anomalocystitidae)
Class Homoiostelea
  Dehmicystis globulus
      (Dehm, 1934) (order Soluta, family
      Dendrocystitidae)
Subphylum Pelmatozoa
Class Blastoidea
(Last systematic treatment Lehmann 1949)

Schizotremites osoleae (Lehmann, 1949)
      (order Fissiculata, family
      Astrocrinidae)
Pentremitidea medusa
      Jaekel, 1895 (order Spiraculata,
      family Troosticrinidae)
Class Rhombifera
(Last systematic treatment Dehm 1932)
  Regulaecystis pleurocystoides Dehm, 1932
      (order Dichoporita, family
      Pleurocystitidae)
Class Crinoidea
(Last systematic treatment Schmidt 1931,
1941)
Subclass Camerata
  Acanthocrinus heroldi
      W.E. Schmidt, 1934 (order
      Diplobathrida, family
      Rhodocriniticae)
  Acanthocrinus lingenbachensis
      Lehmann, 1939 (order
      Diplobathrida, family
      Rhodocrinitidae)
  Acanthocrinus rex
      Jaekel, 1895 (order Diplobathrida,
      family Rhodocrinitidae)
  Arthroacanthaca? claviger
      W.E. Schmidt, 1934 (order
      Monobathrida, family
      Hexacrinitidae)
  Ctenocrinus gracilis
      Jaekel, 1895 (order Monobathrida,
      family Melocrinitidae)
  Ctenocrinus malcontractus
      W.E. Schmidt, 1934 (order
      Monobathrida, family
      Melocrinitidae)
  Culicocrinus spinatus
      Jaekel, 1895 (order Monobathrida,
      family Hapalocrinidae)
  Diamenocrinus opitzi
      W.E. Schmidt, 1934 (order
      Diplobathrida, family
      Rhodocrinitidae)
  Diamenocrinus stellatus
      Jaekel, 1895 (order Diplobathrida,
      family Rhodocrinitidae)

Hapalocrinus elegans
Jaekel, 1895 (order Monobathrida,
family Hapalocrinidae)
Hapalocrinus frechi imbellis
W.E. Schmidt, 1941 (order
Monobathrida, family
Hapalocrinidae)
Hapalocrinus frechi nimisfurcata
W.E. Schmidt, 1934
(order Monobathrida, family
Hapalocrinidae)
Hapalocrinus frechi rarefurcata
W.E. Schmidt, 1941
(order Monobathrida, family
Hapalocrinidae)
Hapalocrinus frechi ssp.
(order Monobathrida, family
Hapalocrinidae)
Hapalocrinus innoxius
W.E. Schmidt, 1934
(order Monobathrida, family
Hapalocrinidae)
Hapalocrinus penninger
W.E. Schmidt, 1934
(order Monobathrida, family
Hapalocrinidae)
Hapalocrinus rauffi
(W.E. Schmidt, 1941)
(order Monobathrida, family
Hapalocrinidae)
Hexacrinus inhospitalis
W.E. Schmidt (order Monobathrida,
family Hapalocrinidae)
Macarocrinus semelfurcatus
W.E. Schmidt, 1941
(order Diplobathrida, family
Dimerocrinitidae)
Macarocrinus springeri
Jaekel, 1895 (order Diplobathrida,
family Dimerocrinitidae)
Macarocrinus terfurcatus
W.E. Schmidt, 1934
(order Diplobathrida, family
Dimerocrinitidae)
Pterinocrinus diensti
W.E. Schmidt, 1934
(order Diplobathrida, family

Dimerocrinitidae)
Pterinocrinus ehrlicheri
W.E. Schmidt, 1934
(order Diplobathrida, family
Dimerocrinitidae)
Thallocrinus acifer
W.E. Schmidt, 1941
(order Monobathrida, family
Hapalocrinidae)
Thallocrinus hauchecornei
Jaekel, 1895 (order Monobathrida,
family Hapalocrinidae)
Thallocrinus procerus
W.E. Schmidt, 1934
(order Monobathrida, family
Hapalocrinidae)
Thallocrinus rugosus
W.E. Schmidt, 1934
(order Monobathrida, family
Hapalocrinidae)
Subclass Disparida
Calycanthocrinus decadactylus lata
W.E. Schmidt, 1934
(family Pisocrinidae)
Calycanthocrinus decadactylus
sp. (family Pisocrinidae)
Senariocrinus maucheri
W.E. Schmidt, 1934
(family Calceocrinidae)
Triacrinus elongatus
Follmann, 1887
(family Pisocrinidae)
Triacrinus koenigswaldi
W.E. Schmidt, 1934
(family Pisocrinidae)
Triacrinus kutscheri
W.E. Schmidt, 1934
(family Pisocrinidae)
Subclass Cladida
Antihomocrinus armatus
W.E. Schmidt, 1934
(family Mastigocrinidae)
Bactrocrinites jaekeli
W.E. Schmidt, 1934
(family Dendrocrinidae)
Bactrocrinites? trabicus
W.E. Schmidt, 1934

(family Dendrocrinidae)

Bathericrinus ericius
 (W.E. Schmidt, 1934)
 (family Mastigocrinidae)

Bathericrinus hystrix
 (W.E. Schmidt, 1934)
 (family Mastigocrinidae)

Bathericrinus semipinnulatus
 (W.E. Schmidt, 1934)
 (family Mastigocrinidae)

Bathericrinus spaciosus
 (W.E. Schmidt, 1934)
 (family Mastigocrinidae)

Dicirrocrinus comtus
 W.E. Schmidt1934
 (order and family uncertain)

Follicrinus grebei
 (Follmann, 1887)
 (family Mastigocrinidae)

Follicrinus kayseri
 (Jaekel, 1895)
 (family Mastigocrinidae)

Gastrocrinus eupelmatus
 (W.E. Schmidt, 1934)
 (family Botryocrinidae)

Gastrocrinus giganteus
 W.E. Schmidt, 1934
 (family Botryocrinidae)

Gissocrinus vertebrachialis
 W.E. Schmidt, 1934
 (family Cyatocrinitidae)

Imitatocrinus gracilior
 (F. Roemer, 1863)
 (family Botryocrinidae)

Iteacrinus dactylus
 (W.E. Schmidt, 1934)
 (family Mastigocrinidae)

Lasiocrinus subramosulus
 W.E. Schmidt, 1934
 (family Dendrocrinidae)

Parisangulocrinus cucumis
 W.E. Schmidt, 1934
 (family Dendrocrinidae)

Parisangulocrinus furcaxialis
 W.E. Schmidt, 1934
 (family Dendrocrinidae)

Parisangulocrinus minax

W.E. Schmidt, 1941
 (family Dendrocrinidae)

Parisangulocrinus schmidti
 Lehmann, 1939
 (family Dendrocrinidae)

Parisangulocrinus zeaeformis
 (Follmann, 1895)
 (family Dendrocrinidae)

Propoteriocrinus scopae
 W.E. Schmidt, 1934
 (family Poteriocrinitidae)

Rhadinocrinus nanus
 (F. Roemer, 1863)
 (family Botryocrinidae)

Rhenocrinus ramosissimus
 W.E. Schmidt, 1934
 (family Rhenocrinidae)

Rhenocrinus lobatus
 Sieverts-Doreck 1973
 (family Rhenocrinidae)

Infraclass Cyathocrinina
Codiacrinus schultzei
 Follmann, 1887
 (family Codiacrinidae)

Infraclass Flexibilia
Eutaxocrinus prognatus
 W.E. Schmidt, 1934
 (order Taxocrinida, family
 Taxocrinidae)

Eutaxocrinus sincerus
 W.E. Schmidt, 1934 (order
 Taxocrinida, family Taxocrinidae)

Taxocrinus stuertzii spinifer
 W.E. Schmidt, 1934 (order
 Taxocrinida, family Taxocrinidae)

Taxocrinus stuertzii
 ssp. Follmann 1887 (order
 Taxocrinida, family Taxocrinidae)

Subphylum Eleutherozoa
Class Edrioasteroidea
(Last systematic treatment Rievers 1961*a*)
 Pyrgocystis (Rhenopyrgus) coronaeformis
 (Rievers, 1961) (order Cyathocystida,
 family Pyrgocystidae)

Class Asteroidea
(Last systematic treatment Lehmann 1957)
 Baliactis devonicus

Spencer, 1922 (order Spinulosida,
family Taeniactinidae)

Baliactis hunsrueckianus
(Lehmann, 1957) (order Spinulosida,
family Taeniactinidae)

Baliactis lancelotus
(Lehmann, 1957) order Spinulosida,
family Taeniactinidae)

Baliactis scutatus
Lehmann, 1957 (order Spinulosida,
family Taeniactinidae)

Baliactis tuberatus
Lehmann, 1957 (order Spinulosida,
family Taeniactinidae)

Compsaster petaliformis
(Stürtz, 1886) (order Forcipulatida,
family Compsasteridae)

Compsaster schlueteri,
(Stürtz 1886) (order Forcipulatida,
family Compsasteridae)

Echinasterella sladeni
Stürtz, 1890 (order Spinulosida,
family Helianthasteridae)

Eostella hunsrueckiana
Lehmann, 1957 (order Valvatida,
superfamily Palaeasteraceae, family
uncertain)

Helianthaster rhenanus microdiscus
Lehmann, 1957 (order Spinulosida,
family Helianthasteridae)

Helianthaster rhenanus
F. Roemer, 1863 (order Spinulosida,
family Helianthasteridae)

Hunsrueckaster peregrinus
Lehmann, 1957 (order Valvatida,
superfamily Palaeasteraceae, family
uncertain)

Hystrigaster horridus
Lehmann 1957 (order Spinulosida,
family Helianthasteridae)

Kyraster inermis
Lehmann, 1957 (order Valvatida,
family Promopalaeasteridae)

Palaeosolaster gregoryi
Stürtz, 1899 (order Paxillosida,
family Palasterinidae)

Palaeostella devonica

(Stürtz, 1886) (order Valvatida,
family Xenasteridae)

Palaeostella solida
Stürtz, 1890 (order Valvatida, family
Xenasteridae)

Palasterina follmanni
Stürtz, 1890 (order Paxillosida,
family Palasterinidae)

Palasterina cornuta
(Lehmann, 1957) (order Paxillosida,
family Palasterinidae)

Palasterina marginata
Lehmann, 1957 (order Paxillosida,
family Palasterinidae)

Palasterina maucheri
Lehmann, 1957 (order Paxillosida,
family Palasterinidae)

Palasterina taenibranchiata
Lehmann, 1957 (order Paxillosida,
family Palasterinidae)

Palasterina tilmanni
Lehmann, 1957 (order Paxillosida,
family Palasterinidae)

Palasteriscus devonicus
Stürtz, 1886 (order Platyasterida,
family Palasterinidae)

Protasteracanthion primus
Stürtz, 1899 (order Forcipulatida,
family Urasterellidae) (the genus *P.* is
assigned to *Urasterella* by Spencer and
Wright (1966), but we regard this as
unlikely)

Urasterella asperula
(F. Roemer, 1863) (order
Forcipulatida, family Urasterellidae)

Urasterella verruculosa
Lehmann, 1957 (order Forcipulatida,
family Urasterellidae)

Class Holothuroidea
(Last systematic treatment Seilacher 1961)
Palaeocucumaria hunsrueckiana Lehmann
1958 (order Arthrochirotida, family
Palaeocucumariidae)

Class Echinoidea
(Last systematic treatment Dehm 1953,
1961 *b*)
?*Lepidocentrus* sp.

(order Echinocystitoidea, family
Lepidocentridae)

Porechinus porosus
Dehm, 1961 (order
Echinocystitoidea, family
Lepidocentridae)

Rhenechinus hopstaetteri
Dehm, 1953 (order
Echinocystitoidea, family
Echinocystidae)

Class Ophiuroidea
(Last systematic treatment Lehmann 1957)

Cheiropteraster giganteus
Stürtz, 1890 (order Lysophiurina,
family Encrinasteridae)

Encrinaster laevidiscus
Lehmann (order Lysophiurina, family
Encrinasteridae)

Encrinaster roemeri
(Schöndorf, 1909) (order
Lysophiurina, family Encrinasteridae)

Eospondylus primigenius
(Stürtz, 1886) (order
Phyrnophiurida, family
Eospondylidae)

Eospondylus primigenius compactus
Lehmann, 1957 (order
Phyrnophiurida, family
Eospondylidae

Euzonosoma opitzi
(Lehmann, 1957) (order
Lysophiurina, family Encrinasteridae)

Euzonosoma tischbeinianum
(F. Roemer, 1863) (order
Lysophiurina, family Encrinasteridae)

Furcaster decheni
(Stürtz, 1886) (order Oegophiurida,
family Furcasteridae)

Furcaster palaeozoicus
Stürtz, 1886 (order Oegophiurida,
family Furcasteridae)

Furcaster zitteli
(Stürtz, 1886) (order Oegophiurida,
family Furcasteridae)

Kenterospondylus decadactylus
Lehmann, 1957 (order
Phyrnophiurida, family

Eospondylidae)

Loriolaster gracilis
Lehmann, 1957 (order Lysophiurina,
family Encrinasteridae)

Loriolaster mirabilis
Stürtz, 1886 (order Lysophiurina,
family Encrinasteridae)

Mastigophiura grandis
Lehmann, 1957 (order
Oegophiurida, family Protasteridae)

Medusaster rhenanus
Stürtz, 1890 (order Stenurida, family
Palaeuridae)

Miospondylus rhenanus
(Stürtz, 1893) (order Oegophiurida,
family Lapworthuridae)

Ophiurina lymani
Stürtz, 1890 (order Oegophiurida,
family Ophiuridae)

Palaeophiura simplex
Stürtz, 1890 (order Oegophiurida,
family Protasteridae)

Stuertzaster (= Erinaceaster) giganteus
Lehmann, 1957 (order Stenurina,
family Pradesuridae)

Stuertzaster (= Erinaceaster) spinosissimus (F.
Roemer, 1886) (order Stenurina,
family Pradesuridae)

Stuertzaster (= Erinaceaster) tenuispinosus
Lehmann, 1957 (order Stenurina,
family Pradesuridae)

Taeniaster beneckei
(Stürtz, 1886) (order Lysophiurina,
family Encrinasteridae)

Taeniaster grandis (Stürtz, 1886) (order
Lysophiurina, family Encrinasteridae)

Phylum Chordata
Class Agnatha
(Last systematic treatment *Drepanaspis* Gross
1963*a*, others Broili 1933*b*)

Drepanaspis gemuendensis
Schlüter, 1887 (order
Psammosteiformes, family
Drepanaspididae)

Pteraspis dunensis smithwoodwardi
Broili, 1933 (order Pteraspidiformes,

family Pteraspididae)

Pteraspis dunensis
.
F. Roemer, 1933 (order
Pteraspidiformes, family
Pteraspididae)

Class Placodermi
(Last systematic treatment *Gemuendaspis*
Westoll and Miles 1963, *Lunaspis* Gross 1961,
Gemuendina Gross 1963*b*, *Tityosteus* Otto
1992, others Broili 1933*b*)

? Brachythoraci gen. et sp. indet.

Gemuendenaspis angusta
(Traquair, 1903) (order Arthrodira,
family Gemuendenaspidae)

Gemuendina stuertzi
Traquair, 1903 (order Rhenanida =
Gemuendinida, family Asterosteidae)

Hunsrueckia problematica
Traquair, 1903 (order Arthrodira,
family uncertain)

Lunaspis broilii
Gross, 1937 (order Petalichthyida,
family Macropetalichthyidae)

Lunaspis heroldi
Broili, 1929 (order Petalichthyida,
family Macropetalichthyidae)

Nessariostoma granulosum
Broili, 1933 (order and family
uncertain)

Paraplesiobatis heinrichsi
Broili, 1933 (order
Pseudopetalichthyida, family
Paraplesiobatidae)

Pseudopetalichys problematicus
Moy-Thomas, 1939 (order
Pseudopetalichthyida, family
Paraplesiobatidae)

Stensioella heintzi
Broili, 1933 (order Stensioellida,
family Stensioellidae)

Stuertzaspis germanica
Traquair, 1902 (order Arthrodira,
family Actinolepidae)

Tityosteus rieversi
Gross, 1960 (order Arthrodira, family
Homostiidae)

Class Acanthodii
(Last systematic treatment Gross 1965)

Machaeracanthus sp. Gross, 1965 (order and
family unknown)

Class Sarcopterygii
(Last systematic treatment Lehmann and
Westoll 1952)

Dipnorhynchus lehmanni
Westoll, 1949 (order Dipnoi, family
Dipnorhynchidae)

Bibliography

This list includes (a) the literature cited in the text, (b) major papers on the paleontology of the Hunsrück Slate published in the last 20 years (i.e. since 1976 – papers published up to that time are listed in the bibliography by Kutscher *et al.* 1980), and (c) the most important older references. The earlier literature on the Hunsrück Slate can be accessed through the major bibliographies of Kutscher *et al.* (1980) and Zwiebelberg (1977). The series of papers by Kutscher (1962–1980) reviewing research on different fossil groups from the Hunsrück Slate includes very comprehensive annotated bibliographies of all the relevant literature.

Ahorner, L. and Murawski, H. (1975). Erdbebentätigkeit und geologischer Werdegang der Hunsrück-Südrand-Störung. *Z. dtsch. geol. Ges. (Hannover)* **126**, 63–82.

Ahrendt, H., Clauer, N., Hunziker, J.C. and Weber, K. (1983). Migration of folding and metamorphism in the Rheinisches Schiefergebirge deduced from K/Ar and Rb/Sr age determinations. In Martin, H. and Eder, F.W. (Eds.) *Intracontinental fold belts.* Springer-Verlag, Berlin, 323–38.

Alberti, G.K.B. (1969). Trilobiten des jüngeren Siluriums, sowie des Unter- und Mitteldevons, I. *Abh. Senckenbergischen Naturforschenden Gesellschaft* **520**, 1–692.

Alberti, G.K.B. (1979). Zur Dacryoconariden-(Tentaculiten-) Chronologie des herzynischen Unter- und Mitteldevons. *Senck. leth.* **60**, 223–41.

Alberti, G.K.B. (1982*a*). Nowakiidae (Dacryoconarida) aus dem Hunsrückschiefer von Bundenbach (Rheinisches Schiefergebirge). *Senck. leth.* **36**, 451–63.

Alberti, G.K.B. (1982*b*). Dacryoconarida from the Lower and Middle Devonian of the Rhenish Schiefergebirge. *Cour. Forschungs. Senckenberg* **55**, 325–32.

Alberti, G.K.B. (1983). Unterdevonische Novakiidae (Dacryoconarida) aus dem Rheinischen Schiefergebirge, aus Oberfranken und aus N-Afrika (Algerien, Marokko). *Senck. leth.* **64**, 295–313.

Allison, P.A. (1990). Pyrite. In Briggs, D.E.G. and Crowther, P.R. *Palaeobiology - a synthesis.*

Blackwell, Oxford, pp. 253–55.

Allison, P.A. and Brett, C.E. (1995). In-situ benthos and paleo-oxygenation in the Middle Cambrian Burgess Shale. British Columbia, Canada. *Geology* **23**, 1079–82.

Allison, P.A. and Briggs, D.E.G. (1991*a*). The taphonomy of soft-bodied animals. In Donovan, S. K. (Ed.) *The Process of Fossilization*, Belhaven, London, pp. 120–40.

Allison, P.A. and Briggs, D.E.G. (1991*b*). Taphonomy of non-mineralized tissues. In Allison, P.A. and Briggs, D.E.G. (Eds.) *Taphonomy: releasing the data locked in the fossil record*, Plenum, New York, pp. 25–70.

Anderle, H.-J. (1987). The Evolution of the south Hunsrück and Taunus Borderzone. *Tectonophysics* **137**, 101–14.

Babcock, L.E. (1991). The enigma of conulariid affinities. In Simonetta, A.M. and Conway Morris, S. (Eds.) *The early evolution of Metazoa and the significance of problematic taxa*. Cambridge University Press, Cambridge, pp. 133–43.

Babcock, L.E. and Feldmann, R.M. (1986). Devonian and Mississippian conulariids of North America. Part A. General description and *Conularia*. *Ann. Carnegie Mus.* **55**, 349–410.

Bale, S.J., Briggs, D.E.G., Parkes, R.J. and Wilby, P.R. (1996). The controls on decay and mineralization – the key to the fossilization of soft tissues. *Sixth N. Am. Paleont. Conv. Abstracts of papers, Paleontological Society Special Publication* **8**, 21.

Bandel, K. and Stanley, G.D. (1989). Reconstruction and biostratinomy of Devonian cephalopods (Lamellorthoceratidae) with unique cameral deposits. *Senck. leth.* **69**, 391–437.

Bandel, K., Reitner, J. and Stürmer, W. (1983). Coleoids from the Lower Devonian Black Slate

('Hunsrück-Schiefer') of the Hunsrück (West Germany). *Neues Jb. Geol. Paläontol.* **165**, 397–417.

Bartels, C. (1985*a*). Dachschieferbergbau in einem Hunsrückdorf. *Gezeiten, Archiv regionaler Lebenswelten zwischen Ems und Elbe (Oldenburg)* **6**, 25–32.

Bartels, C. (1985*b*). Schieferdörfer im Linksrheingebiet. Ein Beitrag zur Geschichte der Arbeit in ländlichen Mittelgebirgsregionen seit dem Ende der Feudalzeit. *Bochumer Archiv für die Geschichte des Widerstandes und der Arbeit* **7**, 7–56.

Bartels, C. (1986). *Schieferdörfer. Dachschieferbergbau im Linksrheingebiet vom Ende des Feudalzeitalters bis zur Weltwirtschaftskrise (1790–1929)*. Centaurus, Pfaffenweiler, 370 pp.

Bartels, C. (1994*a*). Weltberühmt: Die 'Bundenbacher Fossilien' des Hunsrückschiefers. *Schiefer Fachverband in Deutschland e.V., Schriftenreihe*, Bonn **3**, 11–85.

Bartels, C. (1994*b*). Zur Geschichte des Dachschieferbergbaus im Mittelhunsrück um Bundenbach. *Schiefer Fachverband in Deutschland e.V., Schriftenreihe (Bonn)* **3**, 91–132.

Bartels, C. (1995). Die unterdevonischen Dachschiefer von Bundenbach. In Weidert, W.K. (Ed.), *Klassische Fundstellen der Paläontologie III*, Korb, pp. 38–55.

Bartels, C. and Blind, W. (1995). Röntgenuntersuchung pyritisch vererzter Fossilien aus dem Hunsrückschiefer (Unter-Devon, Rheinisches Schiefergebirge). *Metalla. Forschungsberichte aus dem Deutschen Bergbau-Museum, Bochum* **2**(2), 79–100.

Bartels, C. and Brassel, G. (1990). *Fossilien im Hunsrückschiefer. Dokumente des Meereslebens im Devon*. Museum Idar-Oberstein Series **7**, Idar Oberstein, 232 pp.

Bartels, C. and Kneidl, V. (1981). Ein Porphyroid in der Schiefergrube Schmiedenberg bei Bundenbach (Hunsrück, Rheinisches Schiefergebirge) und seine stratigraphische Bedeutung. *Geol. Jb. Hessen (Wiesbaden)* **109**, 23–36.

Bartels, C. and Wuttke, M. (1994). Fossile überlieferung von Weichkorperstrukturen und ihre Genese im Hunrückschiefer (Unter-Ems, Rheinisches Schiefergebirge): ein Forschungsbericht. *Giessener Geol. Schriften* **51** (Festschrift Blind), 25–61, 329–33 (Nachtrag).

Bassler, R.S. (1939). The Hederelloidea, a suborder of Paleozoic Cyclostomatons. *Proc. U.S. Natl Mus.* **87**, 25–91.

Becker, G. and Weigelt, H. (1975). Neue Nachweise von Ophiuroidea im Rheinischen Schiefergebirge. *Notizbl. hess. L.-Amt. Bodenforsch.* **103**, 5–36.

Behr, H.J. (1978). Subfluenz-Prozesse im Grundgebirgs-Stockwerk Mitteleuropas. *Z. dtsch. geol. Ges. (Hannover)* **129**, 283–318.

Benton, M.J. (Ed.) (1993). *The Fossil Record 2*, London, 845 pp.

Bergström, J. (1973). Organization, life and systematics of trilobites. *Fossils and Strata* **2**, 69 pp.

Bergström, J. and Brassel, G. (1984). Legs in the trilobite *Rhenops* from the Lower Devonian Hunsrück Slate. *Lethaia* **17**, 67–72.

Bergström, J., Briggs, D.E.G., Dahl, E., Rolfe, W.D.I. and Stürmer, W. (1987). *Nahecaris stuertzi*, a phyllocarid crustacean from the Lower Devonian Hunsrück Slate. *Paläont. Z.* **61**, 273–98.

Bergström, J., Briggs, D.E.G., Dahl, E., Rolfe, W.D.I. and Stürmer, W. (1989). Rare phyllocarid crustaceans from the Devonian Hunsrück slate. *Paläont. Z.* **63**, 319–33.

Bergström, J., Stürmer, W. and Winter, G. (1980). *Palaeoisopus, Palaeopantopus* and *Palaeothea*, pycnogonid arthropods from the Lower Devonian Hunsrück Slate, West Germany. *Paläont. Z.* **54**, 7–54.

Berthelsen, A. (1992). Mobile Europe. In Blundell, D., Freeman, R. and Mueller, S. (Eds.) *A continent revealed: the European geotraverse.* Cambridge University Press, Cambridge, pp. 11–32.

Birenheide, R. (1971). Beobachtungen am 'Scheinstern' *Mimetaster* aus dem Hunsrück-Schiefer. *Senck. leth.* **52**, 77–91.

Blake, D.B. (1987). A classification and phylogeny of post-Paleozoic sea stars (Asteroidea, Echinodermata). *J. Nat. Hist.* **21**, 481–528.

Blake, D.B. (1994) Reevaluation of the Palasteriscidae Gregory, 1900, and the early phylogeny of the Asteroidea (Echinodermata). *J. Paleont.* **68**, 123–34.

Blake, D.B. and Guensburg, T.E. (1988). The water vascular system and functional morphology of Paleozoic asteroids. *Lethaia* **21**, 189–206.

Blake, D.B. and Guensburg, T.E. (1994). Predation by the Ordovician asteroid *Promopalaeaster* on a pelecypod. *Lethaia* **27**, 235–9.

Blind, W. (1969). Die systematische Stellung der Tentakuliten. *Palaeontographica* **A133**, 101–45.

Blind, W. and Stürmer, W. (1977). *Viriatellina fuchsi* Kutscher (Tentaculoidea) mit Sipho and Fangarmen. *Neues Jb. Geol. Paläontol., Mh.* 513–22.

Blundell, D., Freeman, R. and Mueller, S. (Eds.) (1993). *A continent revealed: the European geotraverse.* Cambridge University Press, Cambridge, 275 pp.

Boucek, B. (1964). *The tentaculites of Bohemia: their morphology, taxonomy, ecology, phylogeny and biostratigraphy.* Czechoslov. Acad. Sci., Prague, 250 pp.

Branco, W. (1906). Die Anwendung der Röntgernstrahlen in der Paläontologie. *Abh. der kgl.-preuss. Akademie der Wissenschaften,* Berlin, pp. 3–55.

Brassel, G. (1968). Wer kauft eine Schiefergrube? Bundenbach-Fossilien werden rar. *Kosmos,* Stuttgart **64**, 363–6.

Brassel, G. (1972). Kaisergrube bleibt erhalten. *Kosmos* **68**, 8–10.

Brassel, G. (1973). So präpariert man Fossilien in Schieferplatten. *Kosmos* **69**, 196–9.

Brassel, G. (1976). Rätselhafte Fährte im Hunsrückschiefer. *Kosmos* **72**, 267–70.

Brassel, G. (1977). Der erste Fund von Hederelloideen (Bryozoa) im Hunsrückschiefer (Unterdevon, Rheinisches Schiefergebirge). *Geol. Jb. Hessen (Wiesbaden)* **105**, 41–5.

Brassel, G. (1985). Die Lebensbedingungen im Hunsrückschiefermeer. *Natur u. Mensch, Jahresmitt. der Naturhist. Gesellsch. Nürnberg* **49**, 75–82.

Brassel, G. (1987). Die Quallen des Hunsrückschiefers. *Natur u. Mensch, Jahresmitt. der Naturhist. Gesellsch. Nürnberg* **51**, 43–6.

Brassel, G. and Bergström, J. (1974). Die Anatomie der Phacopiden des Hunsrückschiefers im Präparat. *Natur und Museum* **104**, 344–9.

Brassel, G. and Bergström, J. (1978). Der Trilobit *Parahomalonotus planus*, ein Homalonotide aus dem unterdevonischen Hunsrückschiefer. *Geol. Jb. Hessen (Wiesbaden)* **106**, 5–11.

Brassel, G. and Kutscher, F. (1978). Beitr. 44: Die Halden der Dachschiefergruben bei Bundenbach (Hunsrück, Rheinisches Schiefergebirge). *Mitt. Pollichia (Bad Dürkheim)* **66**, 11–24.

Brassel, G., Kutscher, F. and Stürmer, W. (1971).

Erste Funde von Weichteilen und Fangarmen bei Tentaculiten. *Abh. hessischen Landesamt. Bodenforsch. (Wiesbaden),* **60** (Heinz Tobien Festschrift), 44–50.

Breimer, A. and Lane, N.G. (1978). Crinoidea: ecology and paleoecology. In *Treatise on invertebrate paleontology,* Part I, Volume 1, Lawrence, Kansas, pp. 316–47.

Brett, C.E. and Seilacher, A. (1991). Fossil Lagerstätten: a taphonomic consequence of event sedimentation. In Einsele, G., Ricken, W., Seilacher, A. (Eds.) *Cycles and events in stratigraphy,* Springer-Verlag, Berlin, pp. 283–97.

Briggs, D.E.G. (1991) Extraordinary fossils. *American Scientist* **79**, 130-141.

Briggs, D.E.G., Bottrell, S.H. and Raiswell, R. (1991). Pyritization of soft-bodied fossils: Beecher's Trilobite Bed, Upper Ordovician. New York State. *Geology* **19**, 1221–4.

Briggs, D.E.G., Erwin, D.H. and Collier, F.J. (1994). *The fossils of the Burgess Shale.* Smithsonian Institution Press, Washington and London, 238 pp.

Briggs, D.E.G. and Kear, A.J. (1993). Decay and preservation of polychaetes: taphonomic thresholds in soft-bodied organisms. *Paleobiology* **19**, 107–135.

Briggs, D.E.G., Kear, A.J., Martill, D.M. and Wilby, P.R. (1993). Phosphatization of soft-tissue in experiments and fossils. *J. Geol. Soc. Lond.* **150**, 1035–8.

Briggs, D.E.G., Raiswell, R., Bottrell, S.H., Hatfield, D. and Bartels, C. (1996). Controls on the pyritization of exceptionally preserved fossils: an analysis of the Lower Devonian Hunsrück Slate of Germany. *Am. J. Sci.* **296**, 633–663.

Broili, F. (1928*a*). Beobachtungen an *Nahecaris. S.B. bayer. Akad. Wiss. math.-nat. Abt. (München)* 1–18.

Broili, F. (1928 *b*). Crustaceenfunde aus dem rheinischen Unterdevon. *S.B. bayer. Akad. Wiss. math.-nat. Abt. (München)* 197–201.

Broili, F. (1928 *c*). Ein Trilobit mit Gliedmassen aus dem Unterdevon der Rheinprovinz. *S.B. bayer. Akad. Wiss. math.-nat. Abt. (München)* 71–82.

Broili, F. (1928 *d*). Ein Pflanzenrest aus den Hunsrückschiefern *S.B. bayer. Akad. Wiss. math.-nat. Abt. (München)* 191.

Broili, F. (1929 *a*). Beobachtungen an neuen Arthropodenfunden aus den Hunsrückschiefern. *S.B. bayer. Akad. Wiss. math.-nat. Abt. (München)* 253–80.

Broili, F. (1929 *b*). Ein neuer Arthropode aus dem rheinischen Unterdevon. *S.B. bayer. Akad. Wiss. math.-nat. Abt. (München)* 135–42.

Broili, F. (1929 *c*). Acanthaspiden aus dem rheinischen Unterdevon. *S.B. bayer. Akad. Wiss. math.-nat. Abt. (München)*, 143–63.

Broili, F. (1930). Ein neuer Nahecaride aus den Hunsrückschiefern. *S.B. bayer. Akad. Wiss. math.-nat. Abt. (München)* 247–51.

Broili, F. (1932 *a*). *Palaeoisopus* ist ein Pantopode. *S.B. bayer. Akad. Wiss. math.-nat. Abt. (München)* 45–60.

Broili, F. (1932 *b*). Ein neuer Crustacee aus dem rheinischen Unterdevon. *S.B. bayer. Akad. Wiss. math.-nat. Abt. (München)* 27–38.

Broili, F. (1933 *a*). Ein zweites Exemplar von *Cheloniellon. S.B. bayer. Akad. Wiss. math.-nat. Abt. (München)* 11–32.

Broili, F. (1933 *b*). Weitere Fischreste aus den Hunsrückschiefern. *S.B. bayer. Akad. Wiss. math.-nat. Abt. (München)* 269–313.

Bromley, R.G. and Ekdale, A.A. (1984). *Chondrites*: a trace fossil indicator of anoxia in sediments. *Science* **224**, 872–4.

Brouwer, A. (1978). Die europäischen Variszciden als Teilstück des nordatlantischen Paläozoikums. *Z. dtsch. geol. Ges. (Hannover)* **129**, 557–63.

Burkhart, E. (1938). Goslars Dachschieferbergbau von seinen Anfängen bis zur Gegenwart. *Beitr. Gesch. Reichsbauernstadt Goslar* **9**, 272 pp.

Burrett, C.F. (1972). Plate Tectonics and the Hercynian Orogeny. *Nature* **239**, 155–7.

Canfield, D.E. and Raiswell, R. (1991). Pyrite formation and fossil preservation. In Allison, P.A. and Briggs, D.E.G. (Eds.) *Taphonomy: releasing the data locked in the fossil record*, Plenum, New York, pp. 337–87.

Caster, K.E. (1967). Homoiostelea. In *Treatise on invertebrate paleontology*, Part 2, Echinodermata 1, Vol. 2, Lawrence, Kansas, pp. 581–627.

Christ, J. (1925). Eine neue fossile Spongiengattung, *Asteriscosella*, im Unterdevon des Nassauischen Hunsrückschiefers; *Asteriscosella nassovica. Jb. Nassau. Ver. Naturk.* **77**, 1–12.

Chlupáč, I. (1976). The oldest goniatite faunas and their stratigraphical significance. *Lethaia* **9**, 303–15.

Conway Morris, S. and Collins, D.H. (1996). Middle Cambrian ctenophores from the Stephen Formation, British Columbia, Canada. *Phil. Trans. R. Soc. Lond.* B **351**, 279–308.

Conway Morris, S. and Whittington, H.B. (1979). The animals of the Burgess Shale. *Scientific American* **241**, 122–33.

Conway Morris, S., Whittington, H.B., Briggs, D.E.G., Hughes, C.P. and Bruton, D.L. (1982). *Atlas of the Burgess Shale*, Palaeontological Association, Nottingham, 31 pp.

Custodis, G. (1990). Schieferabbau und -verarbeitung in Kaub. In Custodis, G.,

Technische Denkmäler in Rheinland-Pfalz, Koblenz, 67–69.

Dallmeyer, R.D., Franke, W., Weber. K. (Eds.) (1995). *Pre-Permian geology of Central and Eastern Europe,* Springer-Verlag, Berlin, 604 pp.

Defrance (1827). Réceptaculite. *Dictionnaire Sci. natur.* **45**, 5–7, Atlas Polypiers, Paris.

Dehm, R. (1932). Cystoideen aus dem rheinischen Unterdevon. *Neues Jb. Geol. Paläontol.,* Stuttgart **B 69**, 63–93.

Dehm, R. (1934). Untersuchungen an Cystoideen des rheinischen Unterdevons. *S.B. bayer. Akad. Wiss. math.-nat. Abt. (München)* 19–43.

Dehm, R. (1953). *Rhenechius hopstätteri* nov. gen. nov. s, ein Seeigel aus dem rheinischen Unter-Devon. *Notizbl. hess. Landesamt. Bodenforsch.(Wiesbaden)* **81**, 88–95.

Dehm, R. (1961*a*). Über *Pyrgocystis* (*Rhenopyrgus* nov. subgen.) *coronaeformis* Rievers aus dem rheinischen Unter-Devon. *Mitt. Bayer. Staatssgl. Pal. hist. Geol. (München)* **1**, 13–17.

Dehm, R. (1961*b*). Ein zweiter Seeigel, *Porechinus porosus* nov. gen. nov. spec. aus dem rheinischen Unter-Devon. *Mitt. Bayer. Staatssgl. Pal. hist. Geol. (München)* **1**, 1–8.

Dehm, R. (1967). Ein weiterer Edrioasteroidee (Echinodermata) aus dem rheinischen Unterdevon. *Mitt. Bayer. Staatssgl. Pal. hist. Geol. (München)* **7**, 175–9.

Dittmar, U. (1996). Profilbilanzierung und Verformungsanalyse im südwestlichen Rheinischen Schiefergebirge. Zur Konfiguration, Deformation und Entwicklungsgeschichte eines passiven variskischen Kontinentalrandes. *Beringia. Würzburger geowissenschaftliche Mitteilungen* **17**, 346 pp.

Dittmar, U., Meyer, W., Oncken, O., Schievenbusch, T., Walter, R. and Winterfeld,

C. v. (1994). Strain partition across a fold and thrust belt: the Rhenish Massif, Mid European Variscides. *J. Struct. Geol.* **16**, 1335–52.

Dittmar, U. and Oncken, O. (1992). Anatomie und Kinematik eines passiven variskischen Kontinentalrandes – zum Strukturbau des südwestlichen Rheinischen Schiefergebirges. *Frankf. Geowiss. Arb.* **A 12**, 34–7.

Dornsiepen, F. (1978). Ein Überblick über die europäischen Varisziden. *Z. dtsch. geol. Ges. (Hannover)* **129**, 521–42.

Durham, J.W., Fell, H.B., Fischer, A.G., Kier, P.M., Melville, R.V., Pawson, D.L. and Wagner, C.D. (1966). Echinoids. In *Treatise on invertebrate paleontology,* Part U, Echinodermata 3, Lawrence, Kansas, pp. 211–640.

Ecke, H.H., Hoffmann, M., Ludewig, B. and Riegel, W. (1985). Ein Inkohlungsprofil durch den südlichen Hunsrück (südwestliches Rheinisches Schiefergebirge), *N. Jb. Geol. Paläont. Mh.* **7**, 395–410.

Engels, B. (1987). Über die Bedeutung der Diagonalstörung im Hunsrückschiefer. *Geol. Jb. Hessen (Wiesbaden)* **115**, 259–84.

Engels, B. and Bank, H. (1954). Ein Querprofil im Bereich der Dachschiefergrube Eschenbach I bei Bundenbach im Hunsrück (Rheinisches Schiefergebirge). *Notizbl. hess. Landesamt. Bodenforsch. (Wiesbaden)* **82**, 247–50.

Erben, H.K. (1953). Goniatitacea (Ceph.) aus dem Unterdevon und dem Unteren Mitteldevon. *N. Jb. Geol. Paläont. Abh.* **98**, 175–225.

Erben, H.K. (1960). Primitive Ammonoidea aus dem Unterdevon Frankreichs und Deutschlands. *Neues Jb. Geol. Paläontol. Abh.* **110**, 1–128.

Erben, H.K. (1962). Zur Analyse und

Interpretation der Rheinischen und Herzynischen Magnafazies des Devons. *Symposium Silur-Devon-Grenze*, Schweizerbarth, Stuttgart, 42–61.

Erben, H.K. (1964). Die Evolution der ältesten Ammonoidea (Lieferung I). *Neues Jb. Geol. Paläontol. Abh.* **120**, 107–212.

Erben, H.K. (1965). Die Evolution der ältesten Ammonoidea (Leiferung II). *Neues Jb. Geol. Paläontol. Abh.* **122**, 275–312.

Erben, H.K. (1994). Das Meer des Hunsrückschiefers. In Koenigswald, W. v. and Meyer, W. (Eds.). *Erdgeschichte im Rheinland. Fossilien und Gesteine aus 400 Millionen Jahren*, Munich, pp. 49–56.

Fauchald, K., Stürmer, W. and Yochelson, E.L. (1986). *Sphenothallus* 'Vermes' in the Early Devonian Hunsrück Slate, West Germany. *Paläont. Z.* **60**, 57–64.

Fauchald, K., Stürmer, W. and Yochelson, E.L. (1988). Two worm-like organisms from the Hunsrück-Slate (Lower Devonian), Southern Germany. *Paläont. Z.* **62**, 205–15.

Fauchald, K. and Yochelson, E.L. (1990*a*). A tubicolous animal from the Hunsrück Slate, West Germany. *Paläont. Z.* **64**, 15–23.

Fauchald, K. and Yochelson, E.L. (1990*b*). Correction: a major error in Fauchald, Stürmer and Yochelson, 1988. *Paläont. Z.* **64**, 381.

Fay, R.O. and Wanner, J. (1967). Blastoids. Systematic descriptions. In *Treatise on invertebrate paleontology*, Part S, Echinodermata 1, Vol. 2, Lawrence, Kansas, pp. 396–455.

Fell, H.B. (1967). Echinoderm ontogeny. In *Treatise on invertebrate paleontology*, Part S, Echinodermáta 1, Volume 1, Lawrence, Kansas, pp. 60–85.

Fischer, W. (1959*a*). Die wirtschaftliche Lage der Bundenbacher Dachschiefergewinnung. *Heimatkalender für den Landkreis Birkenfeld (Neuwied, Rhein)* 113–8.

Fischer, W. (1959*b*). Rheinischer Dachschiefer auf dem Hunsrück. Festbuch für den 12. *Landesverbandstag des Dachdeckerhandwerks Rheinland-Pfalz*, Idar Oberstein, 1–8.

Fischer, W. (1970). Der Dachschieferbau im Hunsrück. *Der Aufschluss*, Sonderband: Idar Oberstein, Heidelberg **19**, 117–28.

Flick, H. and Struve, W. (1984). Beiträge zur Kenntnis der Phacopina (Trilobita) 11: *Chotecops sollei* und *Chotecops ferdinandi* aus devonischen Schiefern des Rheinischen Schiefergebirges. *Senck. leth.* **65**, 137–63.

Follmann, O. (1887). Unterdevonische Crinoiden. *Verh. naturh. Ver. preuss. Rheinl. (Bonn)* **44**, 117–73.

Fougeroux de Bondaroy, A.D. (1763). *Die Kunst den Schiefer aus den Steinbrüchen zu brechen, ihn zu spalten und zu schneiden.* (Original French edition, Paris 1762), Berlin.

Franke, W. (1989). Tectonostratigraphic units in the Variscan belt of central Europe. *Geol. Soc. Am. Spec. Pap.* **230**, 67–90.

Franke, W. (1995). Rhenohercynian foldbelt. Autochthon and nonmetamorphic nappe units: stratigraphy. In Dallmeyer, R.D., Franke, W., Weber. K. (Eds.) *Pre-Permian geology of Central and Eastern Europe*, Springer-Verlag, Berlin, pp. 33–49.

Franke, W., Dallmeyer, R.D. and Weber, K. (1995). Geodynamic evolution. In Dallmeyer, R.D., Franke, W., Weber. K. (Eds.) *Pre-Permian geology of Central and Eastern Europe*, Springer-Verlag, Berlin, pp. 579–93.

Franke, W., Eder, W., Engel, W. and Langenstrassen, F. (1978). Main aspects of geosynclinal sedimentation in the Rhenohercynian zone. *Z. dtsch. geol. Ges. (Hannover)* **129**.

Franke, W. and Oncken, O. (1990). Geodynamic evolution of the North-Central-Variscides – a comic strip. In Freeman, R., Iese,

P. and Müller, S. (Eds.) *The European geotraverse: integrative studies*, European Science Foundation, Strasbourg, pp. 187–94.

Freckmann, K. and Wierschem, F. (1982). *Schiefer – Schutz und Ornament*. Rheinland Verlag, Köln.

Frizzell, D. and Exline, H. (1966). Holothuroidea – fossil record. In *Treatise on invertebrate paleontology*, Part U, Echinodermata, Vol. 3(2), Lawrence, Kansas, pp. 646–72.

Frölich, H. (1924). *Ein Lob der Heimat.* Rheinische Heimatblätter, 358–60.

Fuchs, A. (1907). Die Stratigraphie des Hunsrückschiefers und der Untercoblenzschichten am Mittelrhein nebst einer Übersicht über die spezielle Gliederung des Unterdevons mittelrheinischer Facies und die Faciesgebiete innerhalb des rheinischen Unterdevons. *Zs. dt. geol. Ges. (Berlin)* **59**, 96–119.

Fuchs, A. (1915). Der Hunsrückschiefer und die Unterkoblenzschichten am Mittelrhein (Loreleigegend). I Teil. Beitrag zur Kenntnis der Hunsrückschiefer- und Unterkoblenz Fauna der Loreleigegend. *Abh. preuss. geol. Landesanst. (Berlin)* N.F. **79**, 80 pp.

Fuchs, A. (1930). Versuche zur Lösung des Hunsrückschieferproblems. *Jb. preuss. geol. Landesanst. (Berlin)* **5**, 231–45.

Fuchs, G. (1971). Faunengemeinschaften und Fazieszonen im Unterdevon der Osteifel als Schlüssel zur Paläogeographie. *Notizbl. hess. Landesamt. Bodenforsch.(Wiesbaden)* **99**, 78–105.

Gale, A.S. (1987). Phylogeny and classification of the Asteroidea (Echinodermata). *Zool. J. Linn. Soc. Lond.* **89**, 107–32.

Gall, J.-C. (1983). *Sedimentationsräume und Lebensbereiche der Erdgeschichte. Eine Einführung in die Paläoökologie.* Springer-Verlag, Berlin, 184 pp.

Giese, P. (1978). Die Krustenstruktur des Variszikums und das Problem der Krustenverkürzung. *Z. dtsch. geol. Ges. (Hannover)* **129**, 513–20.

Gilliland, P.M. (1993). The skeletal morphology, systematics and evolutionary history of holothurians. *Spec. Pap. Palaeont.* **47**, 147 pp.

Goldring, R. and Seilacher, A. (1971). Limulid undertracks and their sedimentological implications. *N. Jb. Geol. Paläont.* Abh. **137**, 422–42.

Gross, W. (1933). Die Wirbeltiere des rheinischen Devons. *Abh. preuss. geol. Landesanst. (Berlin)*, N.F. **154**, 1–83.

Gross, W. (1937). Die Wirbeltiere des rheinischen Devons, Part II. *Abh. preuss. geol. Landesanst. (Berlin)*, N.F. **176**, 1–83.

Gross, W. (1949). Die paläontologische und stratigraphische Bedeutung der Wirbeltierfaunen des Old Reds und der marinen altpaläozoischen Schichten. *Abh. der deutschen Akademie der Wissenschaften, mathemat.-naturw. Klasse*, Berlin, 1–130.

Gross, W. (1960). *Tityosteus* n. gen., ein Riesenarthrodire aus dem rheinischen Unterdevon. *Paläont. Z.*, Stuttgart **34**, 263–74.

Gross, W. (1961). *Lunaspis broilii* und *Lunaspis heroldi* aus dem Hunsrückschiefer. *Notizbl. hess. Landesamt. Bodenforsch. (Wiesbaden)* **89**, 17–43.

Gross, W. (1962). Neuuntersuchung der Stensöellidae. *Notizbl. hess. Landesamt. Bodenforsch. (Wiesbaden)* **90**, 48–86.

Gross, W. (1963a). *Drepanaspis gemuendensis* Schlüter. Neuuntersuchung. *Palaeontographica* **A121**, 133–55.

Gross, W. (1963b). *Gemuendina stuertzi* Traquair. *Notizbl. hess. Landesamt. Bodenforsch. (Wiesbaden)* **91**, 36–73.

Gross, W. (1965). Über einen neuen Schädelrest von *Stensiöella heintzi* und Schuppen von *Machaerancanthus* sp. indet. aus dem Hunsrückschiefer. *Notizbl. hess. Landesamt. Bodenforsch. (Wiesbaden)* **93**, 7–18.

Gürich, G. (1931). *Mimaster hexagonalis*, ein neuer Kruster aus dem unterdevonischen Bundenbacher Dachschiefer. *Paläont. Z.* **13**, 204–38.

Gürich, G. (1932). *Mimetaster* n. gen. (Crust.) statt *Mimetaster* Gürich. *Senckenbergiana* **14**, 193.

Haarmann, E. (1922). Die Botryocriniden und Lophocriniden des rheinischen Devons. *Jb. preuss. geol. Landesanst. (Berlin)* **41** (für 1920), 1–87.

Haas, W. (1994). Die Devonischen Riffe der Rheinischen Schiefergebirges im weltweiten und europäischen Rahmen. In Koenigswald, R.v. and Meyer, W. (Eds.) *Erdgeschichte im Rheinland. Fossilien und Gesteine aus 400 Millionen Jahren*, Munich, pp. 71–80.

Hambloch, A. (1913). Die Steinindustrie der Voreifel. In *Eifel-Festschrift zur 25 jährigen Jubelfeier des Eifelvereins*. Bonn, 295–308.

Harland, W.B., Armstrong, R.L., Cox, A.V., Craig, L.E., Smith, A.G. and Smith, D.G. (1990). *A geologic time scale – 1989.* Cambridge University Press, Cambridge, 263 pp.

Hashagen, J. (1913). Zur Geschichte der Eisenindustrie vornehmlich in der westlichen Eifel. In *Eifel-Festschrift zur 25-jährigen Jubelfeier des Eifelvereins*, Bonn, 269–94.

Haude, R. (1995). Die Holothurien-Konstruktion: Evolutionsmodell und ältester Fossilbericht. *N. Jb. Geol. Paläont. Abh. (Festschrift A. Seilacher)*, 181–98.

Heintz, A. (1932). Über einige Fischreste aus dem Hunsrückschiefer. *Zbl. Miner. Geol. Paläont. (Stuttgart)* **B**, 572–80.

Henningsen, D. (1969). Paläogeographische Probleme der Mitteldeutschen Schwelle. *Z. dtsch. geol. Ges. (Hannover)* **121**, 143–50.

Hergarten, B. (1985). Die Conularien des Rheinischen Devons. *Senck. leth.* **66**, 269–97.

Hergarten, B. (1988). Conularien in Deutschland. *Der Aufschluss* **39**, 321–56.

Hergarten, B. (1994). Conularien des Hunsrückschiefers (Unter-Devon). *Senck. leth.* **74**, 273–90.

Herrgesell, G. (1978). *Geologische Untersuchungen im Raume Gemünden, Hunsrück (Rheinisches Schiefergebirge)*. Freiburg i.B. (Diplomarbeit unveröff), 94 pp.

Hirmer, M. (1930). Über ein zweites in den Hunsrückschiefern gefundenes Stück von *Maucheria gemündensis* Broili. *Sitzungsber. d. bayr. Akad. d. Wiss. Math. nat. Abt.*, 39–46.

Holtz, S. (1969). Beitr. 25: Sporen im Hunsrückschiefer des Wispertales (Rheingaukreis, Hessen). *Notizbl. hess. Landesamt. Bodenforsch. (Wiesbaden)* **97**, 389–90.

Houbrick, R.S., Stürmer, W. and Yochelson, E.L. (1988). Rare Mollusca from the Lower Devonian Hunsrück-Slate of Southern Germany. *Lethaia* **21**, 395–402.

Hudson, J.D. (1982). Pyrite in ammonite-bearing shales from the Jurassic of England and Germany. *Sedimentology* **29**, 639–67.

Jaekel, O. (1895). Beiträge zur Kenntnis der palaeozoischen Crinoiden Deutschlands. *Palaeontologische Abh.*, Jena, N.F. **3,1** (Vol. 7 of the whole series), 1–176.

Jaekel, O. (1921). Über einen neuen Phyllocariden aus dem Unterdevon der Bundenbacher Dachschiefer. *Z. dtsch. geol. Ges. (Hannover)* **72** (1920), 290–2.

Jefferies, R.P.S. (1968). The phylum Calcichordata (Jefferies 1967) – primitive fossil chordates with echinoderm affinities. *Bull.*

Brit. Mus. Nat. Hist. (Geol.) **16**, 243–339.

Jefferies, R.P.S. (1984). Locomotion, shape, ornament an external ontogeny in some mitrate calcichordates. *Journal of Vertebrate Paleontology* **4**, 292–319.

Jefferies, R.P.S. (1986). *The ancestry of the vertebrates.* Natural History Museum, London.

Jefferies, R.P.S. (1990). The solute *Dendrocystoides scoticus* from the Upper Ordovician of Scotland and the ancestry of chordates and echinoderms. *Palaeontology* **33**, 631–79.

Jefferies, R.P.S. (1997). A defence of the calcichordates. *Lethaia* **30**, 1–10.

Kaiser, H., Papproth, E. and Stadler, G. (1978). Neue Beobachtungen zur Entstehung des Rheinischen Schiefergebirges. *Z. dtsch. geol. Ges. (Hannover)* **129**, 181–93.

Karathanasopoulus, S. (1975). *Die Sporenvergesellschaftungen in den Dachschiefern des Hunsrücks (Rheinisches Schiefergebirge, Deutschland) und ihre Aussage zur Stratigraphie.* Diss. Univers. Mainz, 96 pp. (unpublished)

Kegel, W. (1950). Sedimentation und Tektonik in der rheinischen Geosynklinale. *Z. dtsch. geol. Ges. (Hannover)* **100** (für 1948), 267–89.

Kesling, R.V. (1967). Cystoids. In *Treatise on invertebrate paleontology*, Part S, Echinodermata 1, Vol. 1, Lawrence, Kansas, pp. 85–267.

Kiepura, M. (1973). Devonian bryozoans of the Holy Cross Mountains, Poland, Pt. II: Cyclostomata and Cystoporata. *Acta Palaeontologica Polonica* **18**, Nr. 4, 323–400.

Kirnbauer, T. (1986). *Geologie, Petrographie und Geochemie der Pyroklastika des unteren Ems, Unter-Devon (Porphyroide) im südlichen Rheinischen Schiefergebirg*e. Diss., Freiburg i. B., 411 pp.

Kirnbauer, T. (1991). Geologie, Petrographie und Geochemie der Pyroklastika des Unteren Ems, Unter-Devon (Porphyroide) im südlichen Rheinischen Schiefergebirge. *Geol. Abh., Hessen (Wiesbaden),* 1–228.

Kirnbauer, T. and Wendorf, K.-W. (1995). Die Fauna der Porphyroide bei Singhofen im Westtaunus (TK 25 Bl. 5713 Katzenelnbogen). *Mainzer geowiss. Mitt.* **24**, 103–54.

Klähn, H. (1929). Die Bedeutung der Seelilien und Seesterne für die Erkennung von Wasserbewegung nach Richtung und Stärke. *Palaeobiologica*, Wien, Leipzig **2**, 287–302.

Kneidl, V. (1980). Zur Geologie des Hunsrücks. *Der Aufschluss* **30**, 87–100.

Kneidl, V. (1984). *Hunsrück und Nahe. Geologie, Mineralogie und Paläontologie. Ein Wegweiser für den Liebhaber.* Kosmos, Stuttgart, 128 pp.

Koch, C. (1880). Über das Vorkommen von *Homalonotus*-Arten in dem rheinischen Unterdevon (4.f: 7. issue) Correspondenz-Blatt, Bonn, 132–141.

Koch, C. (1883). *Homalonotus*-Arten des rheinischen Unterdevons. *Abh. zur geologischen Special-Karte Preussen* Berlin, **4** (2).

Koenigswald, R.v. and Meyer, W. (Eds.) (1994) *Erdgeschichte im Rheinland. Fossilien und Gesteine aus 400 Millionen Jahren.* Dr. Friedrich Pfeil Verlag, Munich, 240 pp.

Koenigswald, R.v. (1930*a*). Die Fauna des Bundenbacher Schiefers in ihren Beziehungen zum Sediment. *Zbl. Miner. Geol. Paläont.* **B**, 241–247.

Koenigswald, R.v. (1930*b*). Die Arten der Einregelung im Sediment bei den Seesternen und Seelilien des unterdevonischen Bundenbacher Schiefers. *Senckenbergiana* **12**, 238–260.

Kott, R. and Wuttke, M. (1987). Untersuchungen zur Morphologie, Paläökologie und Taphonomie von *Retifungus rudens* Rietschel 1970 aus dem Hunsrückschiefer (Bundesrepublik Deutschland). *Geol. Jb. Hessen (Wiesbaden)* **115**, 11–27.

Kräusel, R. and Weyland, H. (1930). Die Flora des deutschen Unterdevons. *Abh. preuss. geol. L.-Anst.*, N.F. **131**, 92 pp.

Krausse, H.F. and Pilger, A. (1976). Betrachtungen zur tektonischen Entwicklung von variszischen Saumsenken in Mittel- und Westeuropa (subvariszische und Catabro-Pyrenäische Saumsenken). *Z. dtsch. geol. Ges. (Hannover)* **127**, 247–69.

Kremer, A. (1951). *Das Dachdeckerhandwerk im Mittelalter, Festschrift anlässlich der Tagung des Landes-Innungs-Verbandes für das Dachdeckerhandwerk Rheinland-Pfalz zu Trier.* Trier, 17–82.

Kuhn, O. (1961). *Die Tierwelt der Bundenbacher Schiefer.* Wittenberg, 56 pp.

Kuhn-Schnyder, E. (1969). Präparation von Fossilien mit dem Sandstrahlgerät. *Naturwissenschaften* **56**, 292–295.

Kutscher, F. (1931). Zur Entstehung des Hunsrückschiefers am Mittelrhein und auf dem Hunsrück. *Jb. Nassau. Ver. Naturk.* **81**, 177–232.

Kutscher, F. (1941). Die Fauna von Berresheim bei Mayen (Bl. Mayen). *Jb. der Reichsstelle für Bodenforsch (Berlin)*, **61** (für 1940), 56–67.

Kutscher, F. (1962*a*). Beiträge zur Sedimentation und Fossilführung des Hunsrückschiefers. *Notizbl. hess. Landesamt. Bodenforsch. (Wiesbaden)* **90**, 160–64.

Kutscher, F. (1962*b*). Beitr. 2: Die Chondriten als Lebensanzeiger. *Notizbl. hess. Landesamt. Bodenforsch. (Wiesbaden)* **90**, 494–8.

Kutscher, F. (1963*a*). Beitr. 3: Die Anwendung der Röntgentechnik zur Diagnostik der Hunsrückschieferfossilien. *Notizbl. hess.*

Landesamt. Bodenforsch. (Wiesbaden) **91**, 74–86.

Kutscher, F. (1963*b*). Beitr. 5: Pteropoden-Vorkommen im Hunsrückschiefer des Hunsrücks und Taunus. *Notizbl. hess. Landesamt. Bodenforsch. (Wiesbaden)* **91**, 366–71.

Kutscher, F. (1964). Beitr. 6: Die Conularien-Arten des Hunsrückschiefers. *Notizbl. hess. Landesamt. Bodenforsch. (Wiesbaden)* **92**, 52–9.

Kutscher, F. (1964*a*). Beitr. 8: *Phacops ferdinandi* Kayser und sein Verbreitungsgebiet in der Hunsrückschieferfazies. *Notizbl. hess. Landesamt. Bodenforsch. (Wiesbaden)* **93**, 19–37.

Kutscher, F. (1965*b*). Beitr. 11: Röhrenbildende Würmer auf Hunsrückschieferfossilien. *Notizbl. hess. Landesamt. Bodenforsch. (Wiesbaden)* **93**, 331–333.

Kutscher, F. (1965*c*). Beitr. 10: Die Vertreter der Klasse Blastoidea (Echinodermata) im Hunsrückschiefer von Kaub und Bundenbach. *Notizbl. hess. Landesamt. Bodenforsch. (Wiesbaden)* **93**, 61–7.

Kutscher, F. (1966*a*). Beitr. 15: *Viriatellina fuchsi* (Kutscher 1931) im Hunsrückschiefer und im Tentaculitenknollenkalk Thüringens. *Paläont. Z.* **40**, 274–6.

Kutscher, F. (1966*b*). Beitr. 13: Lamellibranchiaten des Hunsrückschiefers. *Notizbl. hess. Landesamt. Bodenforsch. (Wiesbaden)* **94**, 27–39.

Kutscher, F. (1966*c*). Beitr. 12: *Acanthocrinus*-Arten im Hunsrückschiefer und im Übrigen Rheinischen Unterdevon. *Notizbl. hess. Landesamt. Bodenforsch. (Wiesbaden)* **94**, 19–26.

Kutscher, F. (1967*a*). Beitr. 17: Ein Orthoceras-Gehäuse mit angehefteten Puellen.

Notizbl. hess. Landesamt. Bodenforsch. (Wiesbaden) **95**, 9–12.

Kutscher, F. (1967b). Beitr. 18: Zur Gattung *Pentremitella* Lehmann 1949. *Notizbl. hess. Landesamt. Bodenforsch. (Wiesbaden)* **95**, 219–20.

Kutscher, F. (1968a). Beitr. 22: Zur Fortführung und Intensivierung der Hunsrückschieferforschung. *Der Aufschluss (Heidelberg)* **19**, 136–9.

Kutscher, F. (1968b). Beitr. 19: Röntgenaufnahmen von Dachschieferplatten mit Tentaculiten, *Jb. Nassau. Ver. Naturk.* **99**, 18–21.

Kutscher, F. (1969a). Beitr. 23: Aus der Frühgeschichte der Untersuchung von Hunsrückschiefer-Fossilien. *Decheniana (Bonn)* **122**, 15–20.

Kutscher, F. (1969b). Beitr. 24: Die Ammonoideen-Entwicklung im Hunsrückschiefer. *Notizbl. hess. Landesamt. Bodenforsch. (Wiesbaden)* **97**, 46–64.

Kutscher, F. (1970a). Beitr. 27: *Palaeopantopus maucheri* Broili und *Palaeoisopus problematicus* Broili. *Notizbl. hess. Landesamt. Bodenforsch. (Wiesbaden)* **98**, 19–29.

Kutscher, F. (1970b). Beitr. 30: Die Echinodermen des Hunsrückschiefer-Meeres. *Abh. hessischen Landesamt. Bodenforsch.* **56**, 37–48.

Kutscher, F. (1970c). 100 Jahre Hunsrückschieferforschung. In *Blatt Mosel, Hochwald, Hunsrück,* Jahrbuch Hunsrückverein 1970, Bernkastel-Kues, 118–125.

Kutscher, F. (1971a). Beitr. 31: Die Verbreitung der Crustaceengattung *Nahecaris* Jaekel im Hunsrückschiefer-Meer. *Notizbl. hess. Landesamt. Bodenforsch. (Wiesbaden)* **99**, 30–42.

Kutscher, F. (1971b). Beitr. 32: *Palaeoscorpius devonicus,* ein devonischer Skorpion. *Jb. d. Nassau. Ver. Naturk. (Wiesbaden)* **101**, 82–8.

Kutscher, F. (1971c). Beitr. 34: Crinoideengrus in einer Dachschieferplatte. *Abh. des hessischen Landesamtes für Bodenforsch.* **60** (Heinz Tobien Festschrift), 113–116.

Kutscher, F. (1972). Beitr. 35: Eine Röntgenaufnahme mit *Bactricrinus jaekeli* W.E. Schmidt und anderen Fossilien. *Mainzer naturwissenschaftliches Archiv* **11**, 83–7.

Kutscher, F. (1973a). Beitr. 37: Zusammenstellung der Agnathen und Fische des Hunsrückschiefer-Meeres. *Notizbl. hess. Landesamt. Bodenforsch. (Wiesbaden)* **101**, 46–79.

Kutscher, F. (1973b). Beitr. 39: Röntgenaufnahmen von Schieferplatten der Gruben Mühlenberg und Schmiedenberg bei Bundenbach. *Jb. Nassau. Ver. Naturk. (Wiesbaden)* **102**, 8–15.

Kutscher, F. (1974a). *Phacops ferdinandi* Kayser 1880, häufigstes Fossil des Hunsrückschiefers. *Blatt Mosel, Hochwald, Hunsrück,* Jahrbuch Hunsrückverein 1974, Bernkastel-Kues, 9–15.

Kutscher, F. (1974b). Beitr, 38: Weitere Arthropodenfunde im Hunsrückschiefer. *Notizbl. hess. Landesamt. Bodenforsch. (Wiesbaden)* **102**, 5–24.

Kutscher, F. (1975a). Beitr. 40: *Rhenopterus diensti,* ein Eurypteride im Hunsrückschiefer. *Notizbl. hess. Landesamt. Bodenforsch. (Wiesbaden)* **103**, 37–42.

Kutscher, F. (1975b). Beitr. 41: 'Cystoideen'-Arten im Hunsrückschiefer. *Geol. Jb. Hessen (Wiesbaden)* **104**, 25–37.

Kutscher, F. (1976a). Beitr. 42: Die Crinoideen-Arten des Hunsrückschiefers. *Geol. Jb. Hessen (Wiesbaden)* **104**, 9–24.

Kutscher, F. (1976b). Beitr. 43: Die Asterozoen des Hunsrückschiefers. *Geol. Jb. Hessen (Wiesbaden)* **104**, 25–37.

Kutscher, F. (1976c). Beitr. 45: Seeigel im Hunsrückschiefer des Hunsrücks mit einem öberblick Über die Echinodermen allgemein. *Jb. Nassau. Ver. Naturk. (Wiesbaden)* **103**, 13–17.

Kutscher, F. (1978a). Die Fauna des Hunsrückschiefers in der Mayener Gegend. *Landesk. Vierteljahresblätter* **24**, Vol. 1, 36–40.

Kutscher, F. (1978b). Beitr. 50: Über Trilobiten des Hunsrückschiefers. *Geol. Jb. Hessen (Wiesbaden)* **106**, 23–52.

Kutscher, F. (1979a). Beitr. 52: *Viriatellina fuchsi* (Kutscher) im Hunsrückschiefer von Gemünden und Bundenbach. *Jb. Nassau. Ver. Naturk.* **104**, 212–18.

Kutscher, F. (1979b). Die Fossilien der Grube Oberer Kreuzberg im Taunus. *Jb. Nassau. Ver. Naturk. (Wiesbaden)* **104**, 206–211.

Kutscher, F. (1979c). Beitr. 51: Gastropoden und Tentaculiten im Hunsrückschiefer. *Geol. Jb. Hessen (Wiesbaden)* **107**, 5–12.

Kutscher, F. (1979d). Beitr. 53: Die Aussagen der Crinoiden Über den Hunsrückschiefer. *Mitt. Pollichia* Bad Dürkheim **67**, 29–43.

Kutscher, F. (1980). Beitr. 54: Spongien im Hunsrückschiefer. *Geol. Jb. Hessen (Wiesbaden)* **108**, 39–42.

Kutscher, F. and Horn, M. (1963). Beitr. 1: Ein Fossilvorkommen im Leimbach-Tal nördlich Bacharach (Unterdevon, Mittelrhein). *Paläont. Z. H. Schmidt Festband*, 134–9.

Kutscher, F., Reichert, H. and Niehuis, M. (1980). *Bibliographie der naturwissenschaftlichen Literatur über den Hunsrück*, Pollichia Buch **1**, Bad Dürkheim, 206 pp. (includes all titles published on the natural history and geology of the Hunsrück region up to and including the year 1976).

Kutscher, F. and Sieverts-Doreck, H. (1968). Beitr. 21: *Pyrgocystis*-Arten im Hunsrückschiefer und mittel- rheinischen

Unterdevon. *Notizbl. hess. Landesamt. Bodenforsch. (Wiesbaden)* **96**, 7–17.

Kutscher, F. and Sieverts-Doreck, H. (1973). Beitr. 36: *Rhenocrinus lobatus* n. s aus dem Hunsrückschiefer. *Notizbl. hess. Landesamt. Bodenforsch. (Wiesbaden)* **101**, 7–15.

Kutscher, F. and Sieverts-Doreck, H. (1977). Beitr. 46: Über Holothurien im Hunsrückschiefer. *Geol. Jb. Hessen (Wiesbaden)* **105**, 47–55.

Larsson, K. (1979) Silurian tentaculitids from Gotland and Scania. *Fossils and Strata* **11**, 180 pp.

Lehmann W.M. (1934). Röntgenuntersuchung von *Asteropyge* sp. Broili aus dem rheinischen Unterdevon. *Neues Jb. Min. Geol. Paläont.* **72**B, 1–14.

Lehmann W.M. (1938). Die Anwendung der Röntgenstrahlen in der Paläontologie. *Jber. Oberrhein. geol. Verein*, n.f. **27**, 16–24.

Lehmann W.M. (1944). *Palaeoscorpius devonicus* n. g. n. sp., ein Skorpion aus dem Rheinischen Unterdevon. *Neues Jb. Min. Geol. Paläont. Mh.* 177–85.

Lehmann W.M. (1949). *Pentremitella osoleae* n.g.n.sp., ein Blastoid aus dem unter- devonis-chen Hunsrückschiefer. *Neues Jb. Min. Geol. Paläont. Mh.* 186–91.

Lehmann W.M. (1955). Beobachtungen und Röntgenuntersuchungen an einigen Crinoiden aus dem Rheinischen Unterdevon. *Neues Jb. Geol. Paläontol.* Abh. **101**, 135–40.

Lehmann W.M. (1956a). Kleine Kostbarkeiten in Dachschiefern. *Der Aufschluss Sonderh.*, Rossdorf bei Darmstadt **3**, 63–74.

Lehmann W.M. (1956b). Beobachtungen an *Weinbergina opitzi* (Merost., Devon). *Senck. leth.* **37**, 67–77.

Lehmann W.M. (1956c). *Dipnorhynchus lehmanni* Westoll, ein primitiver Lungenfisch aus dem rheinischen Unterdevon. *Paläont. Z.* **30**, 21–5.

Lehmann W.M. (1957). Die Asterozoen in den Dachschiefern des rheinischen Unterdevons. *Abh. hess Landesamt. Bodenforsch. (Wiesbaden)*, **21**, 160 pp., 55 pls.

Lehmann W.M. (1958*a*). Eine Holothurie zusammen mit *Palaenectria devonica* und einem Brachiopoden in den devonischen Dachschiefern des Hunsrücks durch Röntgenstrahlen entdeckt. *Notizbl. hess. Landesamt. Bodenforsch. (Wiesbaden)* **86**, 81–6.

Lehmann W.M. (1958*b*). Über einen 21 armigen *Medusaster rhenanus* Stuertz aus dem unterdevonischen Hunsrückschiefer. *Notizbl. hess. Landesamt. Bodenforsch. (Wiesbaden)* **86**, 79–80.

Lehmann W.M. (1959). Neue Entdeckungen an *Palaeoisopus*. *Paläont. Z.* **33**, 96–103.

Lehmann, W. and Westoll, T.S. (1952). A primitive dipnoian fish from the Lower Devonian of Germany. *Proc. R. Soc. Lond.* (B) 140, pp. 403–21.

Liebering, W. (1883). *Beschreibung des Bergreviers Coblenz* I, Bonn, 114 pp.

Maisey, J.G. (1996). *Discovering fossil fishes.* Henry Holt, New York, 223 pp.

Martill, D.M. (1990). Macromolecular resolution of fossilized muscle tissues from an elopomorph fish. *Nature* **346**, 171–2.

Massonne, H.-J (1995). Rhenohercynian foldbelt. Metamorphic units (Northern Phyllite Zone)–metamorphic evolution. In Dallmeyer, R.D., Franke, W., Weber, K. (Eds.), *Pre-Permian geology of Central and Eastern Europe*, Springer-Verlag, Berlin, pp 132–7.

Mehl, D., Wuttke, M. and Kott, R. (1997). Beiträge zur Spongien-Fauna des Hunsrückschiefers (II) Beschreibung eines neuen Kieselschwammes (Hexactinellida,

'Rossellimorpha', fam., gen. et sp. indet.). *N. jb. Geol. Paläont. Mh.* **1997**, 79-92.

Meyer, W. (1965). Gliederung und Altersstellung des Unterdevons südlich der Siegener Hauptüberschiebung in der Südost-Eifel und im Westerwald (Rheinisches Schiefergebirge). In *Max Richter Festschrift*, Clausthal-Zellerfeld, 35–47.

Meyer, W. (1988). *Geologie der Eifel.* 2. erg. Auflage, Stuttgart, 616 pp.

Meyer, W. and Stets. J. (1980). Zur Paläogeographie von Unter- und Mitteldevon im westlichen und zentralen Rheinischen Schiefergebirge. *Z. dtsch. geol. Ges. (Hannover)* **131**, 725–51.

Miles, R.S. (1962). *Gemuendaspis* n. gen., an arthrodiran fish from the Lower Devonian Hunsrückschiefer of Germany. *Trans. R. Soc. Edinb.* **65**, 59–77.

Mittmeyer, H.-G. (1973). Die Hunsrückschiefer-Fauna des Wisper-Gebietes im Taunus (Ulmen Gruppe, tiefes Unter-Ems, Rheinisches Schiefergebirge). *Notizbl. hess. Landesamt. Bodenforsch. (Wiesbaden)* **102**, 16–45.

Mittmeyer, H.-G. (1974). Zur Neufassung der rheinischen Unterdevon-Stufen. *Mainzer geowissenschaftliche Mitteilungen* **3**, 69–79.

Mittmeyer, H.-G. (1978). *Erläuterungen zur Geologischen Karte von Hessen 1:250000*, Blatt Nr. 5813 Nastätten *(Wiesbaden)*, 112 pp.

Mittmeyer, H.-G. (1980*a*). Zur Geologie des Hunsrückschiefers. *Kleine Senckenberg-Reihe*, Frankfurt a.M.**11**, 26–33.

Mittmeyer, H.-G. (1980*b*). Vorläufige Gesamtliste der Hunsrückschiefer-Fossilien. *Kleine Senckenberg-Reihe* **11**, 34–9.

Mittmeyer, H.-G. (1982). Rhenish Lower Devonian stratigraphy. *Cour. Forsch.-Inst. Senckenberg* **55**, 257–70.

Mosebach. R. (1952*a*). Zur Petrographie der

Dachschiefer des Hunsrück-Schiefers. *Z. dtsch. geol. Ges. (Hannover)* **103**, 368–76.

Mosebach. R. (1952*b*). Mineralbildungsvorgänge als Ursache des Erhaltungszustandes der Fossilien des Hunsrück-Schiefers. *Paläont. Z.* **25**, 127–38.

Mosebach. R. (1954). Zur petrographischen Kenntnis devonischer Dachschiefer. *Notizbl. hess. Landesamt. Bodenforsch. (Wiesbaden)* **82**, 234–46.

Moy-Thomas, J.A. and Miles, R.S. (1971) *Palaeozoic fishes.* Chapman and Hall, London, 259 pp.

Müller, K.J. and Walossek, D. (1986). Arthropod larvae from the Upper Cambrian of Sweden. *Trans. R. Soc. Edinb.: Earth Sci.* **77**, 157-179.

Müller, K.J. and Walossek, D. (1988). External morphology and larval development of the Upper Cambrian maxillopod *Bredocaris admirabilis. Fossils and Strata* **23**, 70 pp.

Murawski, H. (1975). Die Grenzzone Hunsrück, Saar-Nahe-Saale-Senke als geologisch- geophysikalisches Problem. *Z. dtsch. geol. Ges. (Hannover)* **126**, 49–62.

Nicolas, A. (1972). Was the Hercynian Orogenic Belt of Europe the Andean type? *Nature* **236**, 221–3.

Nöring, F.K. (1939). Das Unterdevon im westlichen Hunsrück. *Abh. preuss. geol. Landesanst. (Berlin)* N.F. **192**, 1–96.

Oberbergamt Rheinland-Pfalz (1963). *Das Oberbergamt Rheinland-Pfalz in Bad Ems und der Bergbau in seinem Bezirk* (Internationale Industriebibliothek 81/176), Berlin/Basel.

Opitz, R. (1930). Über das Präparieren von Versteinerungen im Hunsrück-Dachschiefer. *Natur und Museum* **59**, 135–40.

Opitz, R. (1931). Seelilien aus den Dachschiefern des Hunsrücks. *Natur und Museum* **60**, 163–8.

Opitz, R. (1932). *Bilder aus der Erdgeschichte des Nahe-Hunsrück-Landes Birkenfeld.* Selbst-Verlag, Birkenfeld, 224 pp.

Opitz, R. (1935). Tektonische Untersuchungen im Bereich des unterdevonischen Dachschiefers südöstlich vom Idarwald (Hunsrück). *Jb. preuss. geol. Landesanst. (Berlin)* **55**, 219–57.

Otto, M. (1992). Ein Neufund des brachythoracen Arthrodiren *Tityosteus rieversi* aus dem unterdevonischen Hunsrückschiefer des rheinischen Schiefergebirges. *N. Jb. Geol. Paläont. Abh.* **187**, 53–82.

Otto, M. (1994). Zur Frage der 'Weichteierhaltung' im Hunsrückschiefer. *Geologica et Palaeontologica*, Marburg **28**, 45–63.

Paul, C.R.C. (1967). The functional morphology and mode of life of the cystoid *Pleurocystites* E. Billings 1854. *Symp. Zool. Soc. Lond.* **20**, 105–23.

Paul, C.R.C. and Smith, A.B. (1984). The early radiation and phylogeny of echinoderms. *Biol. Rev.* **59**, 443–81.

Perroud, H., Van der Voo, R., Bonhommet, N. (1984). Paleozoic geodynamic evolution of the Armorican plate on the basis of paleomagnetic data. *Geology* **12**, 579–82.

Quiring, H. (1926). Die stratigraphische Stellung des Hunsrückschiefers. *Geol. Rdsch* **17a**, 99–109.

Quiring, H. (1931). Römischer Ursprung des Dachschieferbergbaus im Rheinland. *Glückauf* **67**, 1437–8.

Quiring, H. (1932). Die älteste Gewinnung und Verwendung von Dachschiefer im Rheinland. *Forsch. Fortschri. dtsch. Wiss. (Berlin)* **8**, 222.

Rauff, H. (1939). *Palaeonectris discoidea*, eine siphonophoride Meduse aus dem rheinischen Unterdevon, nebst Bemerkungen zur umstrittenen *Brooksella rhenana* Kinkelin. *Paläont. Z.* **21**, 194–213.

Regnell, G. (1966). Edrioasteroids. In *Treatise on invertebrate paleontology*, Part U, Echinodermata 3, Vol. 1, Lawrence, Kansas, pp. 136–73.

Richter, R. (1931). Tierwelt und Umwelt im Hunsrückschiefer; zur Entstehung eines schwarzen Schlammsteins. *Senckenbergiana* **13**, 299–342.

Richter, R. (1935). Marken und Spuren im Hunsrückschiefer 1. Gefliessmarken. *Senckenbergiana* **17**, 244–63.

Richter, R. (1936). Marken und Spuren im Hunsrückschiefer 2. Schichtung und Grundleben. *Senckenbergiana* **18**, 215–44.

Richter, R. (1937). Von Bau und Leben der Trilobiten 8. Die 'Salter'sche Einbettung' als Folge und Kennzeichen des Häutungs-Vorgangs. *Senckenbergiana* **19**, 413–31.

Richter, R. (1941). Marken und Spuren im Hunsrückschiefer 3. Fährten als Zeugnisse des Lebens auf dem Meeresgrunde. *Senckenbergiana* **23**, 218–60.

Richter, R. (1954). Marken und Spuren im Hunsrückschiefer 4. Marken von Schaumblasen als Kennmal des Auftauchbereichs im Hunsrückschiefer- Meer. *Senck. leth.* **35**, 101–6.

Rickard, D. (1996). Fast pyrite formation. In Bottrell, S. (Ed.) *Proc. 4th Int. Symp. Geochemistry Earth's Surface*, 1996. University of Leeds, Leeds, pp.155–9.

Rietschel, S. (1969). Die Receptaculiten. Eine Studie zur Morphologie, Organisation, Ökologie und Überlieferung einer problema-tischen Fossil-Gruppe und die Deutung ihrer Stellung im System. *Senck. leth.* **50**, 465–517.

Rietschel, S. (1970). Beitr. 28: *Retifungus rudens* n.g., n.s, ein dictyospongiider Kieselschwamm aus dem Hunsrückschiefer.

Notizbl. hess. Landesamt. Bodenforsch. (Wiesbaden) **98**, 30–5.

Rietschel, S. and Nitecki, M.H. (1984). Ordovician receptaculitid algae from Burma. *Palaeontology* **27**, 415–20.

Rievers, J. (1961*a*). Eine neuer *Pyrgocystis* (Echinod., Edrioasteroidea) aus dem Bundenbacher Dachschiefer (Devon). *Mitt. Bayer. Staatssammlungen, Pal. hist. Geol. (München)* **1**, 9–11.

Rievers, J. (1961*b*). Zur Entstehung des Bundenbacher Dachschiefers und seiner Versteinerungen. *Mitt. Bayer. Staatssmmlungen, Pal. hist. Geol. (München)* **1**, 19–23.

Ristedt, H. (1968). Zur Revision der Orthoceratidae. *Akademie der Wissenschaften und Literatur, Abh. der mathemat.-naturwiss. Klasse (Mainz) Jg.* **4**, 213–87.

Röder, J. (1970). *Die mineralischen Baustoffe der römischen Zeit im Rheinland.* Bonner Universitätsblätter, 7–19.

Roemer, C.F. (1863). Asteriden und Crinoiden von Bundenbach. *Verh. naturh. Ver. preuss. Rheinl.* **20**, Corr. Bl., 109.

Roemer, C.F. (1862/65). Neue Asteriden und Crinoiden aus devonischem Dachschiefer von Bundenbach bei Birkenfeld. *Palaeontographica (Kassel)* **9**, 143–52.

Roschig, F. (1992). Bibliographie des Dachschiefers (Schwerpunkt Geologie, Bergbau nach ca. 1950 - ausser Thüringen). *Schriftenreihe des Schiefer-Fachverbandes e. V.*, Bonn, Vol. 1.

Ruedemann, R. (1916). Note on *Parapsonema cryptophya* Clarke and *Discophyllum* Hall. *Bull. N.Y. State Mus.* **189**, 22–7.

Runnegar, B. (1980). Hyolitha: status of the phylum. *Lethaia* **13**, 21–5.

Runzheimer, H. (1932). *Novakia gemündina* n.s, ein Pteropode aus dem Hunsrückschiefer (Unterdevon) des Rheinischen

Schiefergebirges. *Senckenbergiana* **34**, 87–91.

Savdra, C.E., Bottjer, D.J. and Gorsline, D.S. (1984). Development of a comprehensive oxygen deficient marine biofacies model: Evidence from Santa Monica, San Pedro, and Santa Barbara Basins, California Continental Borderland. *Bull. Amer. Ass. Petrol. Geol.* **68**, 1179–92.

Schaal, S. and Ziegler, W. (1988). *Messel – ein Schaufenster in die Geschichte der Erde und des Lebens*. Verlag Waldemar Kramer, Frankfurt am Main, 315 pp. (English translation published in 1992 by Oxford University Press, Oxford.)

Schmid, R. (1976). Septal pores in *Prototaxites*, an enigmatic plant. *Science* **191**, 287–8.

Schmidt, W. (1952). Die paläogeographische Entwicklung des linksrheinischen Schiefergebirges vom Kambrium bis zum Oberkarbon. *Z. dtsch. geol. Ges. (Hannover)* **103**, 151–77.

Schmidt, W.E. (1934). Die Crinoideen des Rheinischen Devons, I. Teil: die Crinoideen des Hunsrückschiefers. *Abh. preuss. geol. Landesanst (Berlin)* N.F. **163**, 149 pp.

Schmidt, W.E. (1941). Die Crinoideen des Rheinischen Devons, II. Teil: A. Nachtrag zu: Die Crinoideen des Hunsrückschiefers, B. Die Crinoideen des Unterdevons bis zur Cultrijugatus Zone (mit Aüschluss des Hunsrückschiefers). *Abh. Reichsst. Bodenf. (Berlin)* N.F. **182**, 15–32.

Schöndorf, F. (1909). Die fossilen Seesterne Nassaus. *Jb. Nassau. Ver. Naturk. (Wiesbaden)* **62**, 7–45.

Schuchert, C. (1914). *Fossilium Catalogus, 1: Animalia, pars 3, Stelleroidea palaeozoica*. Junk, Berlin, 53 pp.

Schwarz, J. (1991). Palynostratigraphie im Unterdevon des östlichen Taunus (Blatt 5716 Oberreifenberg und Blatt 5717 Bad Homburg vor der Höhe). *Geol. Abh. Hessen (Wiesbaden)* **93**, 67–81.

Seilacher, A. (1959). Vom Leben der Trilobiten. *Naturwiss.* **46**, 389–93.

Seilacher, A. (1960). Strömungsanzeichen im Hunsrückschiefer. *Notizbl. hess. L.-Amt Bodenforsch* **88**, 88–106.

Seilacher, A. (1961). Holothurien im Hunsrückschiefer (Unterdevon). *Notizbl. hess. Landesamt. Bodenforsch. (Wiesbaden)* **89**, 66–72.

Seilacher, A. (1962). Form und Funktion des Trilobiten-Daktylus. *Paläont. Z.*, H. Schmidt Festband, 218–227.

Seilacher, A. (1970). Fossil-Lagerstätten, Nr. 1, Begriff und Bedeutung der Fossil-Lagerstätten. *Neues Jb. Geol. Paläontol. Mh.* 34–43.

Seilacher, A. and Hemleben, C. (1966). Beitr. 14: Spurenfauna und Bildungstiefe des Hunsrückschiefers. *Notizbl. hess. Landesamt. Bodenforsch. (Wiesbaden)* **94**, 40–53.

Seilacher, A. and Seilacher, E. (1994). Bivalvian trace fossils: a lesson from actuopaleontology. *Cour. Forsch. Senck.* **169**, 5–15.

Selden, P.A. and Jeram, A.J. (1989). Palaeophysiology of terrestrialization in the Chelicerata. *Trans. R. Soc. Edinb.: Earth Sci.* **80**, 303–310.

Simms, M.J. and Sevastopulo, G.D. (1993). The origin of articulate crinoids. *Palaeontology* **36**, 91–109.

Simpson, S. (1940). Das Devon der Südost-Eifel zwischen Nette und Alf. Stratigraphie und Tektonik mit einem Beitrag zur Hunsrückschiefer- Frage. *Abh. Senckenb. naturf. Ges.* **447**, 1–68.

Simpson, S. (1957) On the trace fossil *Chondrites*. *Quart. J. Geol. Soc. London* **112**, 475–99.

Smith, A.B. (1984) Classification of the Echinodermata. *Palaeontology* **27**, 431–59.

Solle, G. (1950). Obere Siegener Schichten, Hunsrückschiefer, tiefes Unterkoblenz und ihre Einstufung ins Rheinische Unterdevon. *Geologisches Jb.* **65**, 299–380.

Solle, G. (1952). Neue Untergattungen und Arten der Bryozoengattung *Hederella* und eine *Hernodia* im rheinischen Unterdevon. *Notizbl. hess. Landesamt. Bodenforsch. (Wiesbaden)* **3**, 35–55.

Solle, G. (1968). Hederelloidea (Cyclostomata) und einige ctenostome Bryozoen aus dem rheinischen Unterdevon. *Abh. des hessischen Landesamtes für Bodenforsch. (Wiesbaden)* **54**, 40 pp.

Solle, G. (1970). Die Hunsrück-Insel im oberen Unterdevon. *Notizbl. hess. Landesamt. Bodenforsch. (Wiesbaden)* **98**, 50–80.

Spencer, W.K. (1934). British Palaeozoic Asterozoa, part 9. *Palaeontograph. Soc. Monog.* **1933**, 437–94.

Spencer, W.K. and Wright, C.W. (1966). Asterozoans. In *Treatise on invertebrate paleontology*, Part U, Echinodermata, Vol. 3(1), Lawrence, Kansas, pp. 4–107.

Stanley, G. jr. and Stürmer W. (1983). The first fossil ctenophore from the Lower Devonian of West Germany. *Nature* **303**, 518–20.

Stanley, G. jr. and Stürmer W. (1987). A new fossil ctenophore discovered by X-rays. *Nature* **327**, 61–3.

Stets, J. (1963). Zur Geologie der Drohntal Schichten und Hunsrückschiefer (Unterdevon) im Gebiet von Bernkastel-Neumagen-Thalfang (Hunsrück, Rheinisches Schiefergebirge). *Notizbl. hess. Landesamt. Bodenforsch. (Wiesbaden)* **90**, 132–59.

Steul, H. (1984). Die systematische Stellung der Conularien. *Giessener Geol. Schriften* **37**, 1–117.

Størmer, L. (1939). Studies on trilobite morphology. Part 1. The thoracic appendages and their phylogenetic significance. *Norsk Geol. Tidskr.* **21**, 49–164.

Struve, W. (1985). Phacopinae aus den Hunsrück-Schiefern (Unterdevon des rheinischen Gebirges). *Senck. leth.* **66**, 393–432.

Struve, W. and Flick, H. (1984). *Chotecops sollei* und *Chotecops ferdinandi* aus den devonischen Schiefern des rheinischen Gebirges. *Senck. leth.* **65**, 137–63.

Stürmer, W. (1968). Einige Beobachtungen an devonischen Fossilien mit Röntgenstrahlen. *Natur und Museum*, Frankfurt a.M. **98**, 413–17.

Stürmer, W. (1969*a*). Röntgenuntersuchungen an paläontologischen Präparaten. *Electromedica*, Erlangen **2**, 48.

Stürmer, W. (1969*b*). Pyrit-Erhaltung von Weichteilen bei devonischen Cephalopoden. *Paläont. Z.* **43**, 10–12.

Stürmer, W. (1970*a*). Die Röntgenaufnahme in der Paläontologie. *Umschau in Wissenschaft und Technik (Frankfurt a.M.)* **18**, 577–88.

Stürmer, W. (1970*b*). Soft parts of Cephalopods and Trilobites: some surprising results of x-ray examination. *Science* **170**, 1300–2.

Stürmer, W. (1974). Röntgenstrahlüng und Paläontologie. *Electromedica* **2**, 43–6.

Stürmer, W. (1980). Portraits from the past: x-rays reveal the intricate beauty of fossils. *Siemens Review* **5**, 16–20.

Stürmer, W. (1984). Interdisciplinary Paleontology. *Interdisciplinary Science Review* **9**, 1–14.

Stürmer, W. (1985). A small coleoid cephalopod with soft parts from the Lower Devonian discovered by using radiography. *Nature* **318**, 53–5.

Stürmer, W. and Bergström, J. (1973). New discoveries on trilobites by x-rays. *Paläont. Z.* **476**, 104–41.

Stürmer, W. and Bergström, J. (1976). The arthropods *Mimetaster* and *Vachonisia* from the Devonian Hunsrück Shale. *Paläont. Z.* **50**, 78–111.

Stürmer, W. and Bergström, J. (1978). The arthropod *Cheloniellon* from the Devonian Hunsrück Slate. *Paläont. Z.* **52**, 57–81.

Stürmer, W. and Bergström, J. (1981). *Weinbergina*, a xiphosuran arthropod from the Devonian Hunsrück Slate. *Paläont. Z.* **55**, 237–55.

Stürmer, W. and Schaarschmidt, F. (1980). Pflanzen im Hunsrückschiefer. *Kleine Senckenberg-Reihe*, Frankfurt a.M. **11**, 19–25.

Stürmer, W., Schaarschmidt, F. and Mittmeyer, H.G. (1980). Versteinertes Leben im Röntgenlicht. *Kleine Senckenberg-Reihe*, Frankfurt a.M. **11**, 80 pp.

Stürtz, B. (1886). Beitrag zur Kenntniss paläozoischer Seesterne. *Palaeontographica* **32**, 75–98.

Südkamp, W.H. (1997). Discovery of soft parts of a fossil brachiopod in the 'Hunsrückschiefer' (Lower Devonian Germany). *Paläont. Z.* **71**, 91–5.

Sutcliffe, O. (1997a) An ophiuroid trackway from the Lower Devonian Hunsrück Slate, Germany. *Lethaia* **30**, 33-9.

Sutcliffe, O. (1997b) *The sedimentology and ichnofauna of the Lower Devonian Hunsrück Slate, Germany: taphonomy and palaeobiological significance.* Unpublished Ph.D. thesis. University of Bristol.

Szudy, K. (1957). Bemerkungen zur Familie Homalonotidae. *Senck. leth.* **38**, 275–90.

Thomson, E. and Vorren, T.O. (1984). Pyritization of tubes and burrows from Late Pleistocene continental shelf sediments of North Norway. *Sedimentology* **31**, 481–92.

Tinnefeld, H. (1989). *100 Jahre Grube Rhein, Bacharach.* Bacharach, 48 pp.

Towe, K.M. (1978) *Tentaculites*: evidence for a brachiopod affinity? *Science* **201**, 626–8.

Traquair, R.H. (1903). The Lower Devonian fishes of Gemünden. *Trans. R. Soc. Edinb.* **40**, 723–39.

Ubaghs, G. (1967a). General Characters of Echinodermata. In *Treatise on invertebrate paleontology*, Part S, Echinodermata 1, Vol. 1, Lawrence, Kansas, pp. 3–60.

Ubaghs, G. (1967b). Stylophora. In *Treatise on invertebrate paleontology*, Part S, Echinodermata 1, Vol. 2, Lawrence, Kansas, pp. 495–565.

Ubaghs, G. (1978). Camerata. In *Treatise on invertebrate paleontology*, Part T, Echinodermata 2, Vol. 2, Lawrence, Kansas, pp. 408–519.

Van der Voo, R. (1988). Palaeozoic palaeo-geography of North America, Gondwana and intervening displaced terranes: comparison of palaeomagnetism with palaeoclimatology and biogeographical patterns. *Bull. Geol. Soc. Am.* **100**, 311–24.

Van Iten, H. (1991). Evolutionary affinities of conulariids. In Simonetta, A.M. and Conway Morris, S. (Eds.) *The early evolution of Metazoa and the significance of problematic taxa.* Cambridge University Press, Cambridge, pp. 145–55.

Van Straelen, V. (1943). *Gilsonicaris rhenanus* nov.gen., nov.sp. branchiopode anostracé de l'Éodévonien du Hunsrück. *Bull. du Musée royal d'Histoire naturelle de Belgique* **19**, 1–10.

Völker, D. (1978). Der Daschieferbergbau in Langhecke. *Nassauische Annalen* **89**, 163–92.

Wagner, W. and Kremb-Wagner, F. (1990). Der Moselschieferbergbau bei Mayen (Rheinisches Schiefergebirge; Unterdevon,

Mayener Dachschieferfolge). *Der Aufschluss* **41**, 202–10.

Walossek, D. (1993). The Upper Cambrian *Rehbachiella* and the phylogeny of Branchiopoda and Crustacea. *Fossils and Strata* **32**, 202 pp.

Weber, K. (1978). Das Bewegungsbild im Rhenoherzynikum – Abbild einer Varistischen Subfluenz. *Z. dtsch. geol. Ges. (Hannover)* **129**, 249–81.

Weber, K. and Behr, H.J. (1987). Geodynamic interpretation of the Variscides. In Martin H. and Eder, F.W. (Eds.) *Intracontinental fold belts.* Springer-Verlag, Berlin, pp. 427–69.

Westoll, T.S. and Miles, R.S. (1963). On an arctolepid fish from Gemünden. *Trans. R. Soc. Edinb.* **65**, 139–153.

Whittington, H.B. (1971). The Burgess Shale: history of research and preservation of fossils. *Proc. N. Am. Paleont. Conv. Chicago 1969*, vol. 1, Allen Press, Lawrence, Kansas, pp. 1170–201.

Whittington, H.B. (1980). The significance of the fauna of the Burgess Shale, Middle Cambrian, British Columbia. *Proc. Geol. Ass.* **91**, 127–48.

Whittington, H.B. (1985). *The Burgess Shale.* Yale University Press, New Haven, London, 151 pp.

Whittington, H.B. (1993). Morphology, anatomy and habits of the Silurian homalonotid trilobite *Trimerus. Mem. Ass. Australas. Palaeontols* **15**, 69–83.

Wignall, P.B. (1994). *Black shales.* Oxford University Press, Oxford.

Wilby, P.R. and Briggs, D.E.G. (1997). Taxonomic trends in the resolution of detail preserved in fossil phosphatized soft tissues. *Geobios, Mémoire spécial*, **20**, 493–502.

Wilby, P.R., Briggs, D.E.G. and Riou, B. (1996). Mineralization of soft bodied invertebrates in a Jurassic metalliferous deposit. *Geology* **24**, 847–50.

Wild, H.W. (1983). Bodenschätze und Bergbau im ehemaligen oldenburgischen Landesteil Birkenfeld. *Schriftenreihe der Kreisvolkshochschule Birkenfeld* **6**, Charivari, Birkenfeld.

Will, V. (1980). Haldenfunde im Hunsrückschiefer. *Der Aufschluss*, (Koblenz) Heidelberg, Special issue **30**, 101–8.

Wills, M.A., Briggs, D.E.G., Fortey, R.A. and Wilkinson, M. (1995). The significance of fossils in understanding arthropod evolution. *Verh. Dtsch. Zool. Ges.* **88**, 203–15.

Winterfeld, C.v. et al. (1994). Krustenstruktur des Rhenohercynischen Falten – und Überschiebungsgürtels. *Göttinger Arb. Geol. Paläont.* Sb.1, 5. Symposium TSK, Göttingen, pp.182–4.

Wolf, M. (1978). Inkohlungsuntersuchungen im Hunsrück (Rheinisches Schiefergebirge). *Z. dtsch. geol. Ges. (Hannover)* **129**, 217–27.

Wuttke, M. (1986). Katalog der Typen und Belegstücke zur Hunsrückschiefer-Sammlung (Rheinisches Schiefergebirge) im Schlossparkmuseum in Bad Kreuznach (BRD). *Mainzer Naturwissenschaftliches Archiv* **24**, 87–117.

Yochelson, E.L. (1989). Reconsideration of possible soft-parts in dacryoconarids (incertae sedis) from the Hunsrückschiefer in western Germany. *Senck. leth.* **69**, 381–90.

Yochelson, E.L., Stürmer, W. and Stanley, G.D. jr. (1983). *Plectodiscus discoideus* Rauff: a redescription of a chondrophorine from the Early Devonian Hunsrück Slate, West Germany. *Paläont. Z.* **57**, 39–68.

Zeiss, A. (1969). Weichteile ectocochleater paläozoischer Cephalopoden in Röntgenaufnahmen und ihre paläontologische Bedeutung. *Paläont. Z.* **43**, 13–27.

Zorn, W. (1967). Neues von der historischen Wirtschaftskarte der Rheinlande. *Rhein. Vierteljahresblätter* **30**, 334–45.

Zschocke, R. (1970). *Die Kulturlandschaft des Hunsrücks und seiner Randlandschaften in ihrer historischen Entwicklung*, Wiesbaden, 240 pp.

Zwiebelberg, W. (1977). *Bibliographie des Hunsrücks* (2nd ed.), Kastellaun, 344 pp. (includes titles on all aspects of the region up to and including the year 1975).

Index